PADDLING WITH A NATURALIST

ALSO BY BRETT HODNETT

HUMAN

PADDLING WITH A NATURALIST

BRETT HODNETT

ISBN 978-1-0689070-2-9

www.bretthodnett.com

Watercolor by Bev Farrer

Nature is a mutable cloud which is always and never the same.

Ralph Waldo Emerson

Preface

A long time ago, a friend of mine agreed to let me take him canoeing one long weekend. A real city dweller, he had never before seen the wilderness firsthand, and had certainly never been in a canoe. After our trip, he said he had enjoyed himself, and that he had felt very manly when we had had to paddle like galley slaves to fight the wind on one particular lake. But he also said that he found the landscape repetitive, just trees and more trees.

A few weeks later, he took me to New York City with him for a week. For me, after just a couple of days, I thought much the same thing, just buildings and more buildings. His comment really made me realize just how much our perceptions and preferences have to do with how much knowledge we have about a subject.

Since this long weekend trip, I have been to university and finished both a bachelor's and a master's degree in biology, and have paddled thousands of kilometers through Canada's lakes and rivers. The more I have learned about the natural world and all of its complexity and interconnectedness, the more fascinated I am with the landscape of "just trees and more trees". Natural selection has developed intricate adaptations in every living thing as each evolves to exist amidst every other living thing. With enough knowledge, just looking at a single leaf on the forest floor can divulge so many fascinating facts, and portray so vividly the interconnectedness of all living things that you can't help but be filled with awe.

I chose to write this book to give canoeists, and anyone else interested in the natural world, a deeper understanding of biology, and to explain why and how the things we see from our canoe are the way they are. I would like to go beyond the identification of species, and the often fascinating, but isolated facts about these species. Instead, I hope to give you a strong sense of the forces behind, and the interconnectedness of this natural world that we are both visitors to and a part of.

To find this beautiful natural world and its inner workings, you don't have to go far. I will be talking to you as I spend a couple of days paddling in Algonquin Park, but it could be anywhere. Over these two days, from the stern of my canoe and from the portage trail, I'll describe and explain what it is that I see in this landscape that is so much more than "just trees". I hope that, like me, with this knowledge you will feel even more awed by the natural world than you already do.

Table of Contents

A New Season

I'm standing on the dock, my gear loaded in the canoe. It's been raining and is overcast, but the air is calm, giving me the feeling that this rainy weather will likely stay for a while. It's the second weekend of May and my first time paddling this season. This winter seemed to defy the changing seasons, keeping the land in its icy grip long after the lengthening days should have passed it into history. Spring has nonetheless finally forced the winter out. The snow has gone and the forest is beginning to come to life. It's still a bit early in the season for most canoeists, and there are none around. The early spring is a time of year so often neglected by canoeists, which is too bad, because this is where it all begins. The leaves are unfurling and many of the plants are in bloom; the birds are all dressed up in their best plumage hoping to find a mate, and if you're lucky, the only mosquitoes and blackflies around are still larvae, inconspicuously hanging out in the ponds and streams.

The lack of people around is an advantage of paddling at this time of year in itself. The lake is quiet, with not a single boat on the water, and the beach is empty. Park staff are repairing the road, and the drone of a front-end loader is the only noise I can hear. Looking out over the quiet lake, it's obvious that it is spring. Although the trees blanketing the hills along shore are mostly coniferous, there are numerous deciduous trees as well. Most of these are still bare, which gives the lake a cold and wintry look. Only the poplars have started to open their leaves. The soft green of their new leaves contrasts with the barren nakedness of the trees around them, warming the scene at least a bit.

With a cursory glance, all of this seems as it should be. We expect the conifers to be green and the deciduous trees to be bare. Then suddenly the bare trees will open up their leaves, just as they have every other year that we can remember. But this scene brings up questions that we often don't think to ask. How do the trees "know" when to open their leaves, and how do they open them so quickly? Why do some trees lose their leaves in the winter while others keep them? And how did the poplars get so lucky as to be the first to open their leaves? Over the next two days as I paddle through lakes and marshes, and portage through wooded trails, nature will pose many questions, and hopefully we can come up with some answers. It will also reveal something of its complexity, as there is often a lot going on behind the scenes that is not at all obvious. Natural selection has created a wide array of complex behaviors and adaptations, as every living thing adapts to find its own niche.

As I push off from shore and take my first paddle stroke of the season, the canoe slides smoothly across the water. I can't think of a better way to travel and see nature. You glide quietly across the water without disturbing anything around you. The canoe allows you to traverse large lakes and explore small streams. If an area can't be paddled, you simply pick your canoe up and carry it to where it can be paddled, exploring the interior of the forests as you go. Perhaps most importantly, the canoe is simple, putting you into a frame of mind in tune with nature, and stripping away all the pretense and unnecessary complication that is our modern life.

The Waking Trees

Once I get over the initial excitement of being in the canoe again, my mind turns back to the trees. Looking around the shore at all the trees, some species with their leaves not yet opened, some species just beginning to open them, one has to wonder how the trees can possibly know when they should begin to open their leaves? We often have warm days in the winter, or even a stretch of unseasonably warm weather in March, only to be followed by snow and freezing temperatures. How do the trees know not to open their leaves? They aren't usually fooled into opening their leaves too early and then having them killed off by cold weather. But on the other hand, it is imperative that they open them as soon as possible, because if they are one of the last trees in the forest to do so, they risk being shaded out by the other trees around them. It is a difficult balance for them to try to find the line where it is as early as possible without being too early. Unfortunately, human-induced climate change has made their job much more difficult. Some regions now have very different weather from what they have had historically, and unseasonable warm and cold spells have become almost commonplace.

There are a number of techniques that plants use to judge when to start their spring growth. The most fundamental of these is to measure the length of the day. By tracking the length of the day, plants can determine what time of year it is, and avoid some of the uncertainty of relying solely on changing weather patterns. Before the days reach a certain length, most plants will not open their leaves to begin their spring growth regardless of the temperature (as an interesting aside, studies have found that it is actually the length of the night that the plants are keeping track of and not the length of the day). By knowing what time of year it is, plants can avoid having a warm spell in early March fool them into opening their leaves too early.

Measuring the length of the days (or nights) is the most effective method of determining the time of year, but it only gives them a rough approximation of when to start growth. It doesn't allow for them to open early if it is an early spring, or to hold off on growth if, like this one, it was a long winter. Since any plant that can find a more accurate way of determining when to start its growth for the season will out-compete the others, it is not surprising that more precise methods have evolved.

A widespread and interesting method that plants use to tune the timing of their spring growth to the weather of a particular season is a system of heat credits. To describe this system, I will give a simplified example of the general system used by many plants.

A particular tree in the forest needs to determine when to begin opening its leaves. The first thing this tree will do is wait for the days to reach a certain length, at which time the tree "knows" that if the conditions are right, it is late enough in the season for it to open its leaves. Once the days have reached this length, this tree will then start to keep track of warm days. In this simplified example, if the average temperature for any particular day is above 6°C, this tree will "count" one point for every degree above. So if the average temperature is 8°C on the first of May, this will count as 2 points (8-6=2). The next day, if the average temperature is 7°C, this will count as one point (7-6=1), and so the grand total will be three points. For any days that are 6°C or below, no new points will be added. The tree will add up its points in this way until it reaches a particular value, we'll say 25, and then its leaves will start to open. However, if at any time before this tree reaches 25 points the temperature falls below a certain value, let's say -3°C, the system resets itself and the tree again starts to count from zero points. This system allows trees to begin spring growth as soon as it is safe to do so, while at the same time ensuring that the cold weather is actually at an end. The numbers I have used are arbitrary, but they illustrate the concept.

Once a tree does decide to open its leaves, they seem to burst open and grow at a remarkable rate. It's amazing that they have enough energy to open all of those thousands of leaves so quickly, especially considering the fact that it is the leaves that use the energy from the sun to create the carbohydrates that they use as their food. It's astonishing, when you think about it, that all the trees in the forest have grown to their present size by using their leaves to build themselves entirely out of the invisible molecules of carbon dioxide in the air.

It would seem that before the leaves are open however, the tree faces a bit of a dilemma. They need food to open their leaves, and leaves to produce their food. The solution, of course, is that the previous fall they stored the sugars produced by their leaves in anticipation of next spring's need. Even the storage of the sugar itself has adaptive significance. These sugars are stored throughout the cells of the tree in the winter where they act as antifreeze, helping the tree to withstand the freezing temperatures. When the winter is over, and the tree no longer needs to protect itself from the freezing temperatures, this antifreeze is then released from the cells and is available to nourish the newly growing leaves. This antifreeze is what we siphon off and boil down to make maple syrup (if the tree happens to be maple). So by storing these sugars, a tree not only aids its survival through the freezing winter, but it also prepares itself to open its buds and quickly produce full-grown leaves to take advantage of the warmer weather when it arrives.

The leaf buds of a tree are themselves well adapted to make full use of these sugars as soon as they are available. Before winter comes and the growing season is fully over, the tree produces its winter buds in preparation for the following spring. If you can manage to take apart and look inside the bud of a tree, you will find a tiny little leaf, already in its characteristic shape, just waiting to grow. This allows the leaves to open quickly since they don't have to undergo any development, all they need to do is increase in size.

Poplars, usually the first to open their leaves, have another trick that allows them to open their leaves quickly in the spring. If you look at the trunk of a poplar, you will see that it has thin bark with a green tinge to it. This bark isn't just incidentally the colour green. The shape, form and function of not just poplars, but of all living things has been selected over time to ensure the survival of that individual and its offspring. So when we ask why something like a poplar tree has a particular characteristic, we can expect that the answer will be because it has some adaptive significance. In the case of poplar trunks, they have a green tinge because they contain chlorophyll. Chlorophyll is the molecule that gives leaves their green colour and that allows photosynthesis to occur, producing carbohydrates, which are the energy source of the plant. So with chlorophyll not only in their leaves but also in their bark, poplars can generate some energy to help them grow and produce flowers even before their leaves have come out. In fact, some photosynthesis occurs in the bark even during the winter. In total up to 15% of their annual carbohydrate gain comes from photosynthesis in their bark, giving them a great advantage when spring first arrives. This is also one of the reasons that poplars, a tree that loses its leaves in the winter, can survive the short growing seasons and long dark winters in the far north.

As I paddle across the lake, I have a great overview of the trees on shore, each one with a long line of ancestors that managed to survive and pass on the skills of calculation and strategy needed to start the season at the right time. One of the nice things about canoeing is that you always have a view. From my location in the middle of the lake, I have a view in four directions. If I were hiking through the forest, I would be very excited about reaching such a view. In the canoe however, I can continue to paddle, almost taking for granted the view that surrounds me.

The paddle is beginning to feel natural in my hand again, and my muscles are recalling the rhythm of the paddle stroke. I can still hear the hum of the front-end loader, and so I feel as though I am paddling away from something and not just paddling for its own sake. Nonetheless, the sound of a single engine seems alone in a vast expanse of quiet. In the city the sound of engines is constant, and nothing is worse in the suburbs than sitting in your yard on a nice day and having to listen to the sound of lawnmowers, weed-whackers and leaf-blowers from every corner of the neighborhood ruining the calm. In our attempt to force nature to behave, we have made our neighborhoods much less livable. By comparison, the lone hum of the front-end loader makes me feel the way that I do on a cool evening when the hordes of mosquitoes have gone away, leaving that single mosquito buzzing around my head. It's an annoyance, but the great sense of relief far outweighs it.

As I approach the opposite side of the lake, more of the details of shore come into view. Like many lakes on the Canadian Shield, the outstanding feature of the shore is that it is made of solid rock. It is difficult for life to get a foothold on a smooth rock surface, but eventually it manages. I am awestruck at the patience of nature and at our inability to think on these same timescales. Often large trees are used as an example of how insignificant our history is in comparison. If you look at the rings of a 150-year-old tree, you can mark off when Canada became a country, the First and Second World Wars, the Berlin Wall coming down, and the beginning of the new millennium. All of this occurred as the tree just sat indifferently in this one spot and grew. And there are trees much older than 150 years. But this is nothing compared to the time it can take for a bare rock surface to become a forest that is able to support a 150-year old tree. During the gradual change from bare rock to forest, a large portion of recorded history has occurred, and all the hubabaloo and self-importance that we attach to the issues of the day seem ridiculously insignificant by comparison.

Life on a Bare Rock: The Lichens

I reach the far shore and paddle along the edge of the lake. Along the shore, I can see many boulders and rocky outcrops where life is just starting to get a foothold. The first organisms that are able to colonize a bare rock are usually the lichens. These diverse and beautiful organisms can live in the harshest of conditions, from hot deserts to arctic tundra, where they spend the vast majority of their time dried out. Only during short periods, when conditions are favorable for growth, will they very slowly spread across the face of a rock or tree. Their tenacity and ability to survive under harsh conditions is explained by the fact that they are not a single organism, but have combined the strengths of two separate Kingdoms, the plants and the fungi. Its main body, which gives the lichen its shape, is a fungus. This fungal body can adhere to a bare surface and absorb water and minerals, withstanding long periods of drought and cold. Nevertheless, on its own, on a bare rock, an individual from this Kingdom would not be able to feed itself. The only Kingdom that can feed itself from almost nothing except the sun are the plants. So the solution that a large variety of fungi have hit upon is to trap a plant, in this case one of a number of species of algae, and embed it in their upper layers exposed to the sun. Here, the algae will do what plants do best. They will utilize the energy of the sun to make carbohydrates out of nothing but water and the carbon dioxide in the air. The lichenized fungi have evolved systems by which they can obtain these carbohydrates from their trapped algal partners. In this way, the algae supplies all the nourishment to the lichen. Since the algae normally makes up less than 5% of the lichen, it is easy to understand why they grow so slowly.

There are over 18,000 different species of fungi that trap a variety of species of algae to make lichens. None of these species of fungi can live without their algal partner, and most of the algae that form this relationship cannot live without their fungal partners. It may occur to you that this creates a dilemma about what an individual is. Can a fungus or an alga that has no chance of living on its own be called an individual? Or is the combination of these two unrelated organisms the individual? When attempting to define a species, we have the same problem. There have been names given to thousands of species of lichens, but when we talk about a species of lichen, what we really mean is a particular combination of two very different "species" that produces a recognizable form. A mysterious thing is that some of the fungi that make up lichens produce spores that are sent off to form new colonies, but these spores are not released with any algal cells, and it is not clear how they hook up with them later.

The Rock Becomes a Forest

Part of the beauty of the Canadian Shield is all the exposed rock that is visible through the vegetation. As I paddle past these rocky shores, it occurs to me that there is actually very little exposed rock that doesn't have some kind of lichen growing on it. Although often overlooked, it is incredible to look at all the colours and patterns of the different lichens, and to imagine where each individual's growth started and how the shapes and patterns came about.

Amongst the lichens covering the bedrock I can see the odd tuft of grass or patch of moss that has managed to find a foothold. Lichens are the first to colonize the bare rock because the fungal partner can withstand very harsh conditions, and the algal partner can produce food for the lichen. The fungi also have the ability to produce acids that break down the rock and allow them to get a better hold on it. In doing this, however, small cracks and crevices in the rock are formed, where sand and organic particles accumulate on what was previously a bare rock surface. This allows some smaller plants, such as grass and mosses, to get a foothold and begin to grow where they otherwise would not have been able to. There is a common misconception that by doing this, lichens are making way for other plants. It is often said or implied that the "goal" of lichens is to produce the conditions so that mosses and grasses can become established, which in turn will create conditions for other plants to become established, before reaching the final "goal" of a full-fledged forest. Nothing could be further from the truth. An individual lichen's only goal is to survive and ensure that as many of its descendants as possible will be around in the future. The lichen has no grand goals of building a forest, and in fact would be better served if the grasses and mosses did not take hold. In fact, lichens will often produce chemicals to try to inhibit the growth of plants. However, unfortunately for the lichen, other organisms are just as eager to survive and take advantage of a situation to the detriment of the lichen. So grasses and mosses slowly gain a foothold, and as they do, more sand and organic material builds up around them. As their roots penetrate into the cracks and crevices prying them apart further, it allows for the establishment of larger plants, much to the chagrin of the mosses and grasses. Eventually, enough soil will accumulate for a forest to grow.

It is a long and arduous path to build a forest from bare rock, and it is amazing that this can take place with no planning at all. This can happen because natural selection gives life an amazing ability to adapt and survive. The process works gradually so that each individual of all the myriad of species finds itself existing in much the same conditions as its ancestors survived in. When a lake trout egg hatches, the new fish will tend to find itself in a deep freshwater lake, and not at the top of a tree in the Kalahari Desert. Fortunately, the individual will also posses the traits and skills that its ancestors relied on to survive those conditions. But each individual will be different in some small random way as well, and those whose differences allow them to take better advantage of the world around them will leave more offspring behind than those without the new difference. The next generation inherits these small random differences, which is the key to why species change over time. Individuals with advantageous traits will become more and more numerous in the population as they out-compete and leave more offspring behind than those individuals without them. Over many generations, all the individuals in the population will have acquired the trait, and the species will be slightly better adapted to its particular way of life than it was before. With individuals from every species competing to survive in this way, the world becomes full of different species, each with its own niche that it has become the expert at surviving in.

As each species struggles to find their niche, they also change the conditions around them, or create new conditions altogether. These new conditions will eventually be taken advantage of by individuals of another species. An intestinal parasite that lives inside of deer would certainly not exist unless the internal environment of deer existed. As different species find and utilize all the conditions that it is possible to survive in, it creates the illusion that there is some greater goal. It is tempting to look at the mosses or small grasses growing on a bare rock as playing their role in the creation of a forest. However, the situation should really be looked at the other way around. These mosses are exploiting an available situation in order to survive, and as they change the conditions around them, individuals of another species will find a way to exploit the new conditions, to the detriment of the mosses.

A Feeling of Wilderness: The Loons

Paddling alongside this complex competition for the right to survive brings me into a narrow bay. As I turn into the bay, the water becomes slightly calmer, and I can no longer hear the drone of the engine across the lake. Now I can feel truly relaxed. It's as though the last layer of tension in the back of my mind has been removed. A loon calls. What better replacement for the sound of an engine than the call of a loon? Does a more beautiful sound exist? Within a couple of minutes, I can see a pair of loons in the water ahead. They allow me to paddle to within about thirty feet of them. They seem particularly bold today. Their pattern of white and black is almost unbelievably distinct and crisp. Before long they grow tired of my intrusion and dive underwater. A minute or so later they appear again, far from the canoe. Now that's a useful adaptation that they have. They've evolved to 'fly' better underwater than in the air. With their legs and feet set far back on their body to propel them through the water, their dense plumage to keep them warm and waterproof, and their bones, denser than most birds, they are ideally suited to a life of diving. Spectacular divers, they have been caught in nets at depths of 80 metres and can remain underwater for over three minutes. Their diving skill allows them to elude potential predators, such as curious canoeists, but it is first and foremost an adaptation to hunt for food. They hunt underwater primarily by sight and so will usually inhabit clear water, like the lake I'm on. Sometimes they will probe around in the vegetation of turbid water, hunting by feel, but their usual lifestyle is to look for fish and chase them down.

Like the trees, loons have also adapted to our long cold winters. As with many other birds, their adaptation is to avoid them altogether and to fly to where it's warmer. These loons likely arrived here shortly after the ice thawed, perhaps returning to the same nest that they left the year before.

Once thought to mate for life, it turns out that loons may change mates, even during the same nesting season if their first nesting attempt fails. Male and female loons look the same, although males tend to be larger than females, and these two probably returned from wintering grounds along the coast, perhaps from as far south as central Mexico. Watching the effort that they put into flying, I can appreciate what a huge undertaking flying south, and then back north each year must be. Because of their heavy bones and the shape of their bodies, loons need a long runway of water to take off and can't actually take off from land.

Sometimes a perfectly healthy loon will become trapped after it lands on a wet road, thinking that it's a river. Once flying, however, they flap their wings quickly and can attain speeds over 120km/h during migration.

I paddle on, leaving the loons to discuss their plans for the summer in peace. I don't get far before a passing cloud pours buckets of rain on me. It only lasts for about five minutes, and then as quickly as it arrived, it has passed, and the rain stops altogether. The passing of the rain has made the bay calm. I take advantage of the calm and stop to eat some lunch. Being alone, I can opt for a lunch that is as simple or as complicated as I please. I opt for simple. A simple lunch allows me to eat in the canoe. I find it interesting to be in the canoe as it floats freely in the water. It tends to drift in unexpected directions, sometimes away from shore, sometimes towards it, often against the wind or current. I've read that Thoreau used to sleep in the bottom of his canoe because he liked the feeling of not knowing where he would be when he got up in the morning.

Ancient Trees and Foamy Water

Sitting in my free-floating canoe, I can hear what sounds like quite a few loons singing and warbling and carrying on. The canoe has drifted against a steep rock face on the northwest shore. Although the steepness of the cliff has kept it mostly bare rock, growing out of the side of the rock face are a number of small cedars. Some of these little trees could be many hundreds of years old. In 1988, researchers from the University of Guelph found that some of the cedars growing out of the side of the cliff on the Niagara Escarpment were more than 800 years old. One dead cedar lived for 1600 years before it finally died. The tiny amount of water and nutrients that they can obtain from their tenuous grip on the cliff face only allows them to grow at a snail's pace. As a result, the trees are still small, despite their age. It is as though the cliff face of the escarpment is a giant bonsai garden.

I spend some time trying to guess the ages of these tiny cedars while I finish my lunch. With the bay still calm, I begin to paddle on my way again. Once I am just around the corner, I can hear the sound of rapids ahead. There is always a sense of anticipation when you can hear, but not see, the rapids ahead of you. It's usually quite easy to get some sense of them from the sound alone. Any set of rapids will make a sound that is something between a gentle rustling noise, like a breeze through the trees, and a throaty roar. These rapids make a sound that is definitely on the "breeze through the trees" end of the scale. Sure enough, as I round the corner, there is a small set of rapids flowing towards me with a little lift-over of about twenty-five or thirty feet on the right shore. In the calm water along shore, a thick white foam covers the water. Although reminiscent of the water coming from a sewage pipe, this foam does not indicate that the water is polluted. The foam is produced when water that is rich in dissolved organic compounds, such as decomposing plants and animals, is aerated by the action of waves or, in this case, a small falls.

Life as a Dragonfly

I reach the lift-over and swing my pack over my back, dragging the canoe the few feet over the rocks to the other side in almost no time at all. Even on this tiny island, just thirty feet of rock, there are grasses, herbs, shrubs, and even a few trees that have managed to establish themselves here. I have a look around the marshy edges on the far side of the portage and come across a dragonfly nymph walking along the muddy bottom in a few inches of water. These somewhat monstrous little creatures are the young stage of the dragonflies canoeists know so well.

Like their parents, these nymphs are predators, eating mosquito larvae and other aquatic invertebrates, and sometimes even small fish. This nymph will have spent the winter in the mud underwater, perhaps not for the first time. Dragonflies spend most of their lives as nymphs and may live their lives as underwater hunters for several years. Dragonflies, in fact, spend very little of their time as adults. Most adult dragonflies live only for a few weeks after spending several years underwater as nymphs. Even the few species of dragonflies that migrate live only long enough to make the flight in one direction. They fly south where they will mate and lay eggs, and it is their offspring that will return the next year.

In the few weeks of adult life that they do have, individual dragonflies compete fiercely with each other to ensure that they are the ones who successfully mate and leave their genes, in the form of offspring, after they die. As a result of this competition, adult dragonflies have evolved complex structural and behavioral adaptations.

Once a female dragonfly has a newly matured batch of eggs that are ready to be fertilized, she will typically look for what she thinks would be a suitable location to leave her offspring. This will be somewhere that her offspring will be able to find food and shelter, and where the water isn't likely to dry up or move too quickly. In other words, a place where her offspring will have the best chance of survival. When choosing a mate, she won't be particularly interested in what the male himself is like. When she arrives at the site, she will typically be mated by whichever male happens to be there. This doesn't mean that she isn't selecting a good mate, however. She is selecting a mate who is strong enough, smart enough or sneaky enough to be at the good egg-laying site when she shows up. This is the main reason that male dragonflies exhibit such strong territorial behavior. It is not uncommon to see one dragonfly chasing another out of his territory, or even to witness a brief scuffle as individuals compete for territory. Usually, a robust male will be able to successfully defend a territory and mate with the females that arrive there. The female gets not only the advantage of a good site to lay her eggs, but also the advantage of mating with a strong, healthy male whose genes will be passed on to her offspring, along with her own, increasing their chance of survival.

Not all males will be able to successfully defend a good quality territory of course, and these individuals will use different strategies. Some may roam from territory to territory looking for an unspoken-for female. Others are sneaks and will hang out near a dominant male's territory and try to intercept females as they approach, or wait for the dominant male to be occupied elsewhere, perhaps already mating, to sneak a mating in with any other females that arrive.

Once a male finds a free female, the competition is far from over. Other males will still try to mate with her, and if she has mated but not yet deposited her eggs into the water, the most recent male to mate with her will try to displace the sperm of previous matings. Different species have evolved different mechanisms to do this. Some have a structurally complex penis with an extendable head or hook-like appendages, which they can use either to scoop out rivals' sperm, or they may push the sperm to a place where its use will be less likely. What males really try to do, however, is to not leave the female available for other matings until she has deposited her eggs.

Males produce their sperm at the end of their abdomen, but in order to make sure that they can keep control of the female during mating, they will actually move it to an organ that is just below their legs. In order for the female to receive the sperm, she has to face the opposite direction to him, forming what is known as the wheel position (don't try this at home). The male has specially adapted appendages at the end of his abdomen that he holds her behind the head with. This allows them to continue flying together and gives him a firm grasp on the back of her neck to prevent other males from mating with her. Some males will hold on to her in this position and dip the end of her abdomen up and down into the water until she has finished washing all of her eggs off. If things are quiet and there aren't many other males around, he may let her go to wash the eggs into the water herself while he hovers nearby and chases off intruders.

Notice that these adaptations are within a species and that natural selection is working on individuals and not the whole species. A male dragonfly must compete for mates with other males of the same species, and those that are successful will be the ones to have their genes and traits passed onto their offspring. It is the species that evolves, but it is because of natural selection acting on individuals that this happens. Hence, over time, because individuals with a particular trait are selected for, the entire species may change completely from what it once was. Or the individuals in a population may be split up and form two separate populations where individuals from one no longer mate with individuals from the other. In this case, the individuals in each population may face different selection pressures, and each population may gradually change until individuals from one population are no longer even capable of mating with individuals from the other population. Now there are two species, each one usually different from their common ancestor.

From the lift-over I can see and hear a pretty little falls ahead. I paddle across to the base of the falls and watch the water as it tumbles down the rocks. Flowing water has a hypnotic quality. Staring at a section of flowing water is much like staring at a campfire. Many hours can go by sitting in front of a fire or water in quiet contemplation.

Whenever I think about the water flowing past, it always astounds me that ultimately all of that water came from the rain. Watching the water in a river pass so quickly by, it is amazing that rivers can flow like this 24 hours a day and not run out of water. Every single drop of that water fell from the sky and somehow slowly made its way through the soil and down tiny creeks to end up flowing in mass down our great rivers. And the rivers continue to flow even though much of the rain doesn't make it to the river. Some of that rain would have ended up in puddles, evaporating back into the air before making it to a river. Much would have been taken up by plants and released back to the atmosphere in that way. Some would seep underground, where it might remain for hundreds of years before being released into a river once again. But eventually it all makes it through, and the rivers continue to flow until they finally empty into the ocean.

Some interesting calculations have been done, which give a vivid example of the interconnectedness of all things on earth and our relationship to water. These calculations estimate what the chances are that in your next glass of water you will ingest a hydrogen atom that was in Julius Caesar's last glass of water. The calculation is quite simple, really. Since almost all the hydrogen in the biosphere is in water (H_2O), when the mass of the hydrogen atoms in Ceasar's 500ml glass of water are divided by the mass of water in the biosphere, and this number is multiplied by the number of molecules in your glass of water, the result is that you would expect there would be 1400 of Caesar's hydrogen atoms in your glass! In other words, there is an essentially 100% chance that the next glass of water you drink will contain at least one hydrogen atom from Caesar's last drink. This holds true for any event that happened long enough ago in the past for the hydrogen atoms to become randomly dispersed around the world. So, a part of every drink of water that you take once flowed as blood through the veins of a Tyrannosaurus rex, and also fell as sweat from the brow of Socrates. The human body consists mostly of water, each of the billions of molecules with its own story to tell.

The Magnificent Leeches

I break the spell of the flowing water and paddle towards the start of the portage trail. As my canoe rides up onto the shore, I see a large leech swimming frantically through the water like a beautiful flying ribbon. When they're moving through the water, and not sucking your blood, leeches are both beautiful and graceful. They also have a remarkable life history. Leeches are hermaphrodites, so each individual is both male and female. They don't self fertilize however, so they must have sex with another leech in order to reproduce. Two leeches will intertwine their bodies, sometimes for several hours, until they eventually line up the spermatophore of one with the clittellar region of the other. The spermatophore is a small sac that the leech will fill with sperm and attach to the body of the other leech near the clitellar region. For any particular mating, the two leeches may both fertilize each other, or one may act only as a male and the other as the female.

Once mating is complete, the sperm will exit the spermatophore and will actually burrow through the skin of the leech into its main body cavity. From there, the sperm will find its way to the ovaries. After the fertilization of the eggs, the leech will secrete a cocoon around its body and fill it with a protein-rich fluid. It will then slide this cocoon off over its head, picking the eggs up along the way. Some species will then deposit the cocoon on a hard surface, such as rock or wood, or for some species onto the back of a crayfish. Many leeches show a surprising amount of parental care. Some species will hover over their cocoon, aerating the water around it to ensure that their young receive enough oxygen while protecting the eggs from predators. Other species will attach their cocoons directly to the bottom of their bodies for the same purpose. During this time, the eggs will hatch, and by the end of two or three weeks the larva will have digested all the protein-rich fluid in the cocoon, and be fully developed. They will then break through the cocoon and attach themselves to their parent, where they will stay until they are delivered to their first meal. The young will then leave the parent to feed, and when they have finished this first meal, they will finally go their own way.

The bloodsucking leeches don't need to eat very often. They can store up to five times their body size in blood, which allows them to go months between meals. Depending on the species of leech, they will have either one or two species of bacteria living inside of their crop and intestine in a symbiotic relationship. In exchange for food and a place to live, these bacteria produce antibiotics, allowing the leech to store blood for long periods of time without it going bad. This blood in the crop is then released into the intestine a little bit at a time to nourish the leech. Leeches are incapable of digesting this blood themselves, however, and they rely on their symbiotic partners to do this for them as well. These bacteria also produce vitamins for the leech, particularly B-complex vitamins, which animals that live on blood typically have a problem obtaining for themselves. Clearly, both the leeches and the bacteria depend entirely on this symbiotic relationship to survive, and one would not be able to live without the other.

It may seem incredible that a symbiotic relationship with bacteria is so essential to leeches, but to biologists, what is actually interesting is the simplicity of the relationship; the fact that they only have one or two species of bacteria in their gut doing this. The situation in the human intestine is more usual. We have between 500 and 1000 different species of microorganisms, the vast majority of which are bacteria, in our intestines. In fact, our total microbial population contains at least ten times as many cells as our body does and accounts for nearly two kilograms of our body weight. These microorganisms perform many functions for us. They break down a number of compounds that our body isn't capable of breaking down, synthesize certain vitamins, as well as improve our immune system. Unlike with the leeches, the full importance of our relationship with these microorganisms is extremely difficult to determine because it is an entire community interacting with one another and not just a species or two with specific functions.

The leech has disappeared from view, and my focus moves to the land. From the canoe, it looks like an interesting little portage. The trail is a small gully that goes up a steep 20-foot hill. The gully is a collage of sharp rocks and tree roots, and on this overcast spring day, there is a steady stream of water running over the rocks, making it appear more like a creek than a portage trail. I stand at the base of this creek and unload the canoe. It is a short portage, but it is long enough that I will strap the paddles in the back and carry the canoe in proper portage fashion.

I have come up with my own personal rating system for different-length portages. A portage of less than 150 m is Class I. The most trying thing about a Class I portage is taking all the stuff out of the canoe, and then repacking it at the other end. There is nothing worse than six or eight Class I portages one after the other.

Class II portages are from 150 m to 750 m in length. These are my favorite portages, long enough that they are worth taking the stuff out of the canoe for. They are a break from paddling and allow you to stretch your legs and get your blood flowing.

Class III portages start to get a little long. I put these in the range of 750 m to 2500 m. Carrying a canoe this far starts to hurt and goes beyond getting your blood flowing and into sweat and discomfort. I generally don’t look forward to Class III portages. They do have one very important positive feature, however. In a region that is visited by large numbers of people, a Class III portage will frighten off 90% of them, leaving you with a more wilderness feel on the other side.

The final class is Class IV. These are the portages that are over 2500 m in length. There is only one possible strategy for these portages, and that is to pretend that they never end. You have to be firmly decided in your mind that you will continue to carry your canoe through the bush for the rest of your life. Once you commit to this, a Class IV portage isn't quite as bad. It took me quite a few of these long portages to figure this out, but if at any point on one of these portages you start to think that the end of the portage must be "just over this hill", you are done for. Because the end of the portage is not just over the hill you are on, and when you get to the next hill, it is not just over that hill either. You need to recognize that a Class IV portage has no end. Looking over each hill for the end will make Dante's descent to hell seem like, well, a Class III portage.

These classes, of course, must be modified in some situations. A short portage through a thigh-deep bog, or up a steep slope where ropes need to be used to pull the canoe to the top qualify as entirely different experiences from your average portage.

This portage that lies ahead of me is only 90m long, putting it in my Class I portage category. I unload the canoe and head up the path. The trail is extremely wet, with deep puddles in every depression. Along the path, leaf litter covers the forest floor. Fresh new spring plants push through dried leaves of oak and cedar. Spring flowers begin to show their colours. Mayflower creeps over the forest floor, its clusters of pink flowers stand out amidst the brown of the fallen leaves. Little purple violets are scattered where the wind has kept the ground free of fallen debris. The shiny new leaves of cinquefoil and meadow rue stretch up towards the sun, and strawberry plants open their leaves to the sky where the ground is drier and the sun shines uninhibited. On the trail with a light drizzle falling, the contrast between last fall's death and this spring's birth is striking. But it's easy to see that last year's dead leaves and plants do not represent death in any morbid or final sense. They are merely a stage in the cycle of the forest. Without the nutrients that they return to the ground, new growth could not occur.

Succession and the Oaks

On the forest floor around me, many of the leaves that cover the ground are oak. Looking around at the trees overhead, this might come as a surprise, since very few are oak. There are a few oak seedlings around, but protected from the wind, these still have most of last year's leaves on them. In fact, this is not uncommon for oaks. Oak trees tend to hold on to their leaves long into the winter, which is part of the reason that their leaves are so abundant in the spring. As well, when they do drop their leaves, the leaves break down very slowly and are therefore on the ground for a long time, sometimes taking a few years before they are unrecognizable as oak leaves. This contrasts sharply with a tree like maple, whose leaves easily break down and are soon unrecognizable. The result is that oak leaves often dominate the forest floor, while the maples are quickly turned into soil.

The question of course arises as to why this should be the case. As with most things in nature, it isn't just happenstance, but has to do with strategy and survival. We have talked about succession, the conversion of bare rock into a forest. But succession also happens within a forest as well. Oak trees have carved out their niche as a tree that grows best on well-drained soils in direct sunlight. When a forest is just emerging, trees like oak and hickory thrive. Many forests in the Great Lakes Region are dominated by oak and hickory and are even given the name oak-hickory forests. But if you take a walk through one of these forests, you will find that there are very few oak trees in the understory. Almost all the saplings in the forest will be of other species, such as maple or beech. This is because the sun-loving oaks can't grow in their own shade, whereas maple and beech thrive in the oak's shade. The mighty oaks, standing tall and proud, may look powerful, dominating their forest kingdom, but below them a peasant revolt is going on. These young beech and maple trees will eventually grow tall, and when the oaks grow old and die, the forest will then be a beech-maple forest. A beech-maple forest is considered a climax forest because now when the seeds of the beech and maples fall, they can successfully grow and replace their parents, and the forest will remain a beech-maple forest.

If you are a stately oak then, what are you to do? When you drop your leaves at the end of the season and they break down to make new soil, the nutrients will not be used by your descendants, but instead will feed the hungry young maple trees. The oaks have come up with the best strategy that they could. They hold on to their leaves for as long as they can, withdrawing as many nutrients from them as possible. This way, the tree can use the nutrients for its own growth and survival, helping it to live a long life and produce large quantities of acorns. With any luck, a squirrel, or maybe a chipmunk or mouse, will carry one of these many acorns to somewhere that the conditions are right for it to grow to maturity, and the process will start again. When the leaves finally do fall, they don't have many nutrients left in them, and most soil microorganisms are not very interested in them. Since they don't have much to offer to the decomposers, the leaves sit around on the forest floor for a long time. A maple tree, by contrast, is more than happy to drop its nutrient-rich leaves on the forest floor, since the nutrients will be taken up by that tree's very own offspring, allowing their genes to be carried on indefinitely.

Violets: What's So Great about Sex

Seemingly oblivious to the battle raging around them, the spring flowers rush to get most of their life-cycle finished before the leaves of whichever trees stand above them fully open overhead and block the sun. These flowers have battles of their own, and wouldn't have time to think about the tree's problems, even if they did have a philosophical bent.

The violets (*Viola sororia*) are already in flower. Violets work hard to ensure that they will produce seeds. During the flowering season, a violet will adjust the position of its flower stalk so that the flower face will have a nodding, then horizontal, and finally erect position. Each position attracts a different kind of pollinator, and so increases the violet's chances of becoming pollinated and successfully producing seeds. If the bees don't successfully pollinate them, why not try the butterflies? If the butterflies don't get the job done, maybe the flies will pollinate them. If all of these attempts to receive pollen from another violet (cross-pollination) are unsuccessful, the violet has a back-up plan. As the season progresses, they will send up shorter and shorter stalks until summer arrives and their flowers look like wide green buds at the base of the plant. These are not buds, however, and they will not open to become a recognizable flower. They are cleistogamous flowers, and if you look below the leaves of a violet in the summer, you can often see them. These flowers, which never open, have both male and female parts (stamens and a pistil) and will pollinate themselves and successfully produce seeds whether insects visit them and they are cross-pollinated or not.

You might wonder why violets don't just produce cleistogamous flowers in the first place. It would clearly save a lot of energy and trouble. The reason for this is that through natural selection, violets have found that the energy spent becoming cross-pollinated is worth the effort because of the advantages of sexual reproduction. In other words, the individual violets that went to the trouble of cross-pollinating tended to leave more successful descendants than those individual violets that did not cross-pollinate, and so the violet population eventually only contained individuals that cross-pollinated.

Sexual reproduction is important because an individual is more likely to have some of its offspring survive if each of its offspring has a variety of different traits and skills, obtained by the recombining of genes from its two parents. These genes, which the parents are passing down, have proven themselves to be successful by the simple fact that the two parents managed to survive and reproduce. Combining the two sets of genes in random ways is what allows most of the variation in populations to come about, and as we talked about earlier, changing to adapt to a changing world is what keeps organisms competitive. The only other source of variation is from mutations, and although mutations are always the original source of variation, the process is very slow. Most of the variation that you find within a species is from the recombining of existing genes rather than from new mutations. Mutations also have another disadvantage, which is that they usually make things worse for an individual rather than better. Chernobyl and other such disasters expose animals, including people, to high levels of radiation, which causes many mutations in these individuals. When this occurs, the results are tragic, and we don't expect these exposed individuals to have stronger, smarter children. The reason that mutations are usually bad is because there is no way an organism can direct where they will occur and what the results will be. It is a totally random process, and only very rarely does it improve things. If you think about how complicated and intertwined the workings of an organism such as a violet or a human being are, it becomes clear that randomly changing something will probably be unsuccessful. But over long periods of time with this process working on large numbers of individuals, because those few advantageous mutations are the ones that get passed on, and the disadvantageous mutations are not, the individuals that are alive today are vast conglomerations of advantageous mutations. Natural selection allows random mutations to drive a completely non-random process.

For the violets, although it is to their advantage to reproduce sexually, if that doesn't work, self-pollinating allows them to produce offspring that will still have a good chance of surviving and becoming successfully cross-pollinated in the next season. It's certainly better than not reproducing at all.

Why Latin? *Epigaea repens*

Taking my attention away from the violets are the most common spring flower along the trail, mayflower or *Epigaea repens*. These trailing clusters of small pink flowers are sometimes also called trailing arbutus. This plant is a good example of why it is usually preferable to use the scientific name when identifying a species rather than one of their common names. Common names can lead to confusion in a number of ways.

The biggest problem is that common names arise independently in different regions, sometimes giving the same plant many different names. *Plantago major*, which is often called broadleaf plantain, is a common plant around the world. Most people would recognize it as the dandelion-like rosette of leaves on their lawn but with a stalk of small greenish-white flowers, rather than the large pretty flower of the dandelion. As a result of the ubiquitous nature of *Plantago major*, besides being called broadleaf plantain, it has 45 other English names, 11 French names, 75 Dutch names, and 106 German names. The opposite happens as well; two or more completely unrelated plants can end up with the exact same name. By using scientific names, the mayflowers along my portage trail will not be confused with another plant that is common in the area, Canada mayflower. Mayflower (*Epigaea repens*) is from the heath family (Ericaceae), whereas Canada mayflower

(*Maianthemum canadense*) is from the lily family (Liliaceae).

Scientific names may at first appear to be a lot more complicated than common names, but overall they are actually simpler. Scientific names use two names to describe each species, the first being the genus and the second the species. A particular species will always be identified using both the genus and species names. The genus name is shared by a group of similar and related species, while the species name identifies the individual species within this group. In the case of *Epigaea repens*, *Epigaea* is the name given to the small group of closely related species, and *repens* is the name given to the specific species that I can see along the trail within that group. Only one species within each genus will have the species name *repens*, even though another genus may also have a species within it that uses the name *repens*. So, it is always necessary to give both genus and species names when identifying a species.

As well as being binomial, genus and species names also use Latin or Latinized words. Latin is used because it is a dead language that is not commonly spoken, and so it is not likely to change much. As well, Latin is the root of many present-day languages, and so it is recognizable to people of many different linguistic backgrounds.

Using the name *Epigaea repens* then, has several advantages over the common name. To begin with, you can be confident that only one species has this name, and that this is the only name that this species has, so it is clear to others which species you are talking about. Secondly, the scientific name can be used and understood when communicating with people of any language. A scientist in China or Sweden can't be expected to know the local Russian common names of all the plants when communicating with their colleague in Russia. Because they all use the Latin binomial however, when their Russian colleague is discussing a species, the foreign scientists will know immediately which plant he is speaking about, or if it is a species which is new to them, they will recognize the genus name and know where the plant fits in the broader picture.

This is another advantage of using Latin binomials. The name itself attempts to tell you something about the evolutionary relationships between species. Common names can be extremely misleading in this respect. An example from another one of our spring flowers is dogtooth violets. Dogtooth violets are not violets at all, but are actually from the lily family. Their scientific name is *Erythronium americanum*, and they are very closely related to both Minnesota adder's tongue (*Erythronium propullans*) and white trout lily (*Erythronium albidum*). This is obvious from their scientific names, where they are all from the same genus, but is far from clear using their common names.

Sex Gets More Complicated: Early Meadow Rue

Along the trail of this short Class I portage, the other plants aren't as fast to get their flowers open as the *Epigea repens* and *Viola sororia*. Instead, their new leaves are busy growing as quickly as they can. Some of these plants are easy to identify by their leaves even before the flowers open. One of these is early meadow rue (*Thalictrum dioicum*). Their tall stems and distinctive leaves at the end of long stalks give early meadow rue a somewhat ungainly yet delicate appearance. It won't be long before their drooping, feathery flowers, held aloft by even finer stalks, will add to the delicate appearance of the plant.

The species name, *dioicum,* not only distinguishes this species from the other species in the genus *Thalictrum*, but it also conveys some important information about this particular species. The name *dioicum* is derived from a Greek word meaning two households, referring to this species' unusual method of flower production. Unlike most plant species, an individual of the species *Thalictrum dioicum* will produce either male or female flowers, but not both. This may not seem like an unusual strategy when we think of animals, but in plants it is quite uncommon.

Any plant species can use one of three strategies when producing flowers. They can produce separate male and female plants like *Thalictrum dioicum* (dioecious), they can produce separate male and female flowers grown on the same plant (monoecious), or they can produce flowers that have both male and female parts (hermaphroditic). Most species of plants with conspicuous flowers are hermaphroditic, having both male and female parts in the same flower. This is the common strategy used by most of the pretty, colourful plants grown in gardens, and also by most of the more striking flowers in the forest. Trilliums, dogtooth violets, mayflowers, Canada mayflowers, violets and most other flowers you come across in the forest have flowers that contain both male and female parts. The main advantage of this is that it allows them to both distribute pollen and to become fertilized with just one visit by a pollinator. This greatly improves their chances of reproductive success. Another advantage is that for most plants, at a certain point in the season, the female flowers, or parts of flowers, stop being receptive to pollen and focus instead on developing their seeds. If you are a male-only plant, you can do nothing further to promote your reproductive success at this time because there are no females around to pollinate. If you are a hermaphrodite, however, you can dispense with the male role and focus your energies on developing your seeds.

Although hermaphroditic plants are quite common, it is misleading to think that most plants have both sexes in one flower just because most of the flowers that we see are this way. Many of our most common plants use the second strategy; they have flowers of different sexes on the same plant (monoecious). Most of these plants, however, have inconspicuous flowers because they are commonly species that are pollinated by the wind rather than by insects or birds. Because these plants don't need to attract pollinators, they typically don't go to the trouble of producing beautiful flowers and instead tend to produce rather dull, inconspicuous flowers that are easily overlooked. Adaptation to wind pollination is usually accompanied by a separation of the male and female parts into separate flowers in order to minimize self-pollination. Many very common plants have both male and female single-sex flowers on the same plant, such as birches, alders, oaks, cattails, sedges and grasses. The conifers, although they don't produce flowers, also have separate male and female cones on the same tree.

It is often precisely because they are so common that these species are wind pollinated. It would be difficult to attract enough pollinators to each of the millions of flowers in a field of grass. But at the same time, because they are so common, it is quite likely that if a plant releases a lot of pollen into the wind, much of it will reach the female flowers of the numerous other plants of the same species that are all around it. Having separate male and female flowers makes it easier for these plants to avoid self-pollination. Not only are the flowers separated in space from each other, but they can also open their male and female flowers at different times, making it impossible for them to self-fertilize.

By having both male and female parts on the same plant, monoecious species also have the advantage that hermaphroditic species have. That is that when the female flowers stop being receptive to pollen they can focus on developing their seeds and let their male flowers die back, unlike a male only plant which would be stuck with nothing left to do to promote their reproductive success after the female plants stopped receiving pollen.

Early meadow rue is one of a smaller number of plants that use the third strategy and have flowers of different sexes on different plants (dioecious). There are a few common plants that use this strategy, including willows and poplars, as well as holly (*Ilex aquifolium*). In some situations, having each plant be a separate sex can be advantageous as well. A practical advantage is that having separate sex plants eliminates the problem of inbreeding. Unlike species with both sexes on the same individual, dioecious plants don't need to ensure that male and female flowers open at separate times, or they don't need to evolve potentially costly mechanisms so that an individual's pollen can't pollinate its own flowers.

Perhaps a more significant advantage of early meadow rue's dioecious strategy is the ability to determine which sex it would be most advantageous for the individual to become, and to then be able to choose that sex. In some cases, it is the female of the dioecious plant who determines what sex its seeds will be, depending upon conditions at the time. If there is plenty of pollen around, and if the environmental conditions seem to be favourable, a female may produce seeds that will become female, since females require more resources and better conditions to be able to go through the whole process of not only producing flowers, but of also then producing seeds. If the conditions do not seem favourable, she may instead produce seeds that will become male plants.

In some species, the sex of a dioecious plant is not predetermined at all. It may be determined instead by environmental conditions, and may change from year to year. So in the first year, when a plant is just becoming established, it may produce only male flowers, and then in its second year, when it has enough resources to develop fruit and seeds, it may produce only female flowers. Or if the plant finds itself in dry, poor conditions, it may remain male after the first year, or revert back to male after several years of being female.

Some species of plants take advantage of all the reproductive strategies. Jack in the pulpit (*Arisaema triphyllum*) is a great example of this. One individual plant may be nonsexual in its first year and then become male in its second year. As it grows bigger, it may produce both male and female flowers, and when it is fully established, it may have almost exclusively female flowers. The ability of plants to adapt is truly astounding.

Jeepers Creepers: Strawberries and Cinquefoil

At the end of the trail there is a clearing, and where the sun is shining and it is a bit drier, there is another plant with distinctive leaves, the common strawberry (*Fragaria virginiana*). Even those not familiar with our wild plants can usually recognize the three-parted leaves whose stalks arise directly from the ground, if only to find the delicious berries that will appear under them later in the summer. The so-called berries of the strawberry are not actually berries at all. The juicy red edible portion is actually the enlarged end of the floral stem, called the receptacle. It is actually each of the little “seeds” on the surface of the receptacle that is a tiny, hard fruit. In contrast, a berry is a fleshy ovary (or several ovaries fused together) each with many seeds. So, oddly enough, in the biological sense, tomatoes, watermelons and grapefruits are berries, but strawberries and raspberries are not.

The leaves of another plant, Canadian dwarf cinquefoil (*Potentilla canadensis*), are also abundant in the clearing at the end of the trail. These plants, like the common strawberry, are in the family Rosaceae or Rose family. As well as producing seeds so that they can reproduce sexually, both plants spread by creeping runners, a form of vegetative reproduction. A parent plant will send out these thin stem-like runners up to half a metre away in every direction, producing little clones of themselves along them.

The parent plant will become encircled by these cloned plantlets, and will send water and nutrients to them until they can establish themselves. This is a very effective way of both colonizing open ground, and of forming a continuous ground cover to crowd out other species attempting to grow in the same area.

I pile my gear back into the canoe, and leave the *Epigaea repens, Thalictrum dioicum, Potentilla canadensis, Fragaria virginiana* and *Viola sororia* to face their challenges on their own. The next portage is only metres away, a short Class II. I paddle a couple of strokes to the other side and pull my gear out of the canoe again. The start of this portage is flat, making it easy to pull the canoe out of the water before carrying it up a fairly steep rocky hill. There are two big old trees at the start of the portage, a white pine and a red pine. Along the trail, the trees are sparse as there is still a lot of exposed rock around. The empty spaces between the trees are pretty, with carpets of lichens and mosses. The trail stays along the top of the hill, and the ground isn't quite as wet as it was on the last portage. I'm not in a hurry, so I can take my time and enjoy my surroundings. There are no mosquitoes or blackflies at all, allowing me to comfortably stop and examine anything that strikes me as being interesting. After carrying my pack to the other end of the portage, I enjoy retracing my steps to get the canoe. Unencumbered, I can take a closer look at anything that caught my eye the first time across, as well as find new things to investigate.

On this portage, the flowers of *Epigaea repens* are again widespread, and along the shore leatherleaf (*Chamaedaphne calyculata*) is in flower. Both of these early bloomers are in the heath family, a family largely made up of shrubs or woody perennial herbs, which are evergreen. Herbaceous plants can lengthen their growing season by being evergreen, just as conifers do. Being close to the ground, a layer of snow usually covers herbaceous plants throughout much of the winter. This protects them and insulates them from the extreme cold. However, being beneath the snow does not necessarily mean that they are not active. Plants under snow have been shown to be undergoing photosynthesis even when the snow is as much as 80cm deep. Although the snow blocks most of the light, the wavelength of light that penetrates to the greatest depths is near the optimum for photosynthesis.

Being evergreen also means that they will already have leaves when the warmth of spring comes, and they can immediately produce flowers while many of the other spring plants are still growing their new seasons' leaves. The flowers of leatherleaf are, in fact, almost fully formed by the previous fall so that they are ready to bloom when spring comes. This gives them an early start to the season, and almost exclusive access to any insect pollinators that are out, helping to ensure that they will successfully reproduce.

Cooperation with the Animal Kingdom: The Pollinators

Having their flowers open so early in the season wouldn't do the Leatherleaf plant much good if it was still too cold for any insects to be active. Any insects that pollinate them must also be adapted to the cool weather of the early spring. Not coincidentally, they are most often pollinated by an insect that is well equipped to survive and stay active in cold weather. Bumblebees (*Bombus spp.*).

Bumblebees have many adaptations that allow them to be active when the weather is still quite cold. Some of these adaptations have to do with their behavior. Bumblebees do not hover, and, as an adaptation to living in cold climates, will often walk from flower to flower allowing their body temperature to drop to below what is necessary for them to fly. When they then need to fly, they will heat their bodies up by contracting their muscles and shivering until they are warm enough to move their wings quickly and create lift. Bumblebee queens also act in this manner to keep their brood warm. Early in the season when it is cold out, a queen will perch on her eggs, larvae, and pupae and press her abdomen onto them like a bird. By shivering and keeping herself warm, she can also keep her young warm.

The body design of bumblebees is itself one of the most fundamental of the ways that bumblebees stay warm. The thick "fur" that covers their bodies is an effective insulation, and the large size of their bodies means they have less surface area than most insects and so do not cool as quickly. Their dark colour helps them to absorb the heat of the sun, while at the same time, by contrasting it with a bright yellow, they can warn predators that it would not be wise to eat them. A toad that eats a bumblebee will never eat a second one.

Another method they use to retain their body heat takes advantage of the design of insects' circulatory and respiratory systems. For all animals, the purpose of breathing is to exchange the carbon dioxide they create as a waste product, with fresh oxygen from the air. To be in anything but a dormant state, an animal can only go a very short time without fresh oxygen. Mammals and birds make this exchange using their lungs, where the gases are then transferred to their circulatory systems-the blood vessels and heart. It is their circulatory system that then transports these gases along with nutrients, water, proteins, hormones and antibodies throughout their body. It is a closed system where the blood always stays within the confines of vessels with discrete veins and arteries.

Insects have a very different circulatory system. Theirs is an open circulatory system where the blood does not stay in vessels. Instead, blood is pumped by the heart into an artery, which ends in cavities where the blood then flows freely around and through the body's tissues. Eventually the blood ends up in a cavity surrounding the heart, where it is pumped throughout their whole body again. As in a closed circulatory system, this allows insects to transport nutrients, water and proteins throughout their bodies. Insects, however, do not use lungs to exchange carbon dioxide for oxygen, and so these gases are not transferred to the circulatory system. Instead, an insect's body is full of little tubules which have openings to the outside through tiny valves. These tubules deliver oxygen and remove carbon dioxide directly throughout their bodies. In some insects this happens passively, but in others, usually larger insects like the bees, the tubules are attached to elastic air sacs which expand and contract to help in the exchange process.

Because oxygen is carried separately from blood throughout an insect's body, they can stop their hearts for long periods without cutting off their oxygen supply. This is what bumblebees do. When it is cold out, they will stop their heart so that the blood stays in their thorax and is not pumped to the abdomen where its heat would be dissipated to the outside air.

It should not be surprising that the pollinators that have adaptations to allow them to be active early in the season, and the flowers that are in bloom at this time, should be adapted to one another. As is so often the case in nature, natural selection has selected for species that have co-operated, and each is dependent on the other for their survival. If the early season pollinators disappeared, the plants that rely on them would not be able to reproduce, and if the plants didn't bloom early enough, the pollinators would starve.

Of course the pollinators of leatherleaf must not only be out at the right time of year, they must also be able to pollinate the type of flower that leatherleaf produces. The flowers of leatherleaf are particularly well adapted to bumblebees. They are like tiny bells hanging along one side of the plant's stem. Flowers such as these are not pollinated easily. They have no landing rim, so any insect that pollinates them must be a good flier, and they require the pollinator to have a long tongue to reach deep inside the floral tube to the nectar glands at the base. Bumblebees are agile fliers, and their long, thin tongues are fully 50% to 85% of their body length, up to twice the length of honeybees, giving them almost exclusive access to the nectar of the flowers of leatherleaf.

There are many species of plants whose flowers have been evolving in concert with specific species of insects for a very long time. Some species of both plants and insects are generalists, taking advantage of a large range of potential pollinators or flower types. Others have built a relationship where their reliance on each other has allowed them to evolve a strategy that benefits both of them. For these plants, it is in their best interest to restrict access to their pollen to a few specialized species of insect. This greatly improves the chances of their pollen being delivered to another plant of the same species where fertilization can take place. They normally do this by designing their flowers in such a way that only a specific pollinator can gain access to them. For example, some species of plant may utilize the fact that bees are agile fliers with long tongues, colour vision and the ability to recognize detailed patterns. A plant can produce flowers that take advantage of these adaptations. The flowers may have a small landing platform with the nectar hidden deep inside a tube. The patterns on the flower make them recognizable and guide the bee to the nectar at the base of the tube. These guides on the flower will be in the colours bees can see, which aren't always visible to the human eye. As well as being able to see all the colours humans can see except for red, bees can also see wavelengths of light that are shorter than the human eye can, such as UV. Many flowers that look plain white or plain red to humans actually have intricate UV patterns that are clearly visible to bees on them.

If these plants were generalists and all insects could easily get to their pollen, their pollen would not be delivered so directly to other flowers of the same species. Most of their pollen would be wasted, as insects delivered it to every species of plant in the neighborhood. From the bee's point of view, if they wish to maximize the amount of pollen that they collect, going to flowers that are very likely not to have been visited by other insects ensures they will be rewarded at almost every flower. In fact, when scout bees return to their hives and do their famous dances, they will indicate to the other bees not only the distance and direction to the location of flowers, but the pollen and scent on the scouts will also indicate the type of flower. When the bees then arrive at the location the scout has designated, they will look specifically for these types of flowers, and will tend to ignore other flowers, even if they are a species that they normally visit.

The flower/pollinator relationship is an excellent example of natural selection favouring species that co-operate with one another. There is an often-mentioned view that survival of the fittest means only the strong survive, and that in nature every individual and every species is simply pitted against every other individual and species. Throughout our recent history, this view has become so ingrained in our psyche that it has been routinely used to justify the oppression of one group of people by another, and is still used to justify having a solely competitive economic system. It has become increasingly apparent to biologists, however, that it isn't just the ability to out-compete each other that brings success. Natural selection seems to select for cooperation between different species, perhaps more often than not. In many cases, one species can't exist without the other. There are many examples of this in nature. We've talked about flowers and their pollinators, bacteria in the intestines of animals, and also about the relationship between fungi and algae to make up lichens. But there are countless others, some of which we will undoubtedly come across.

The Astonishing Mosses

Retracing my steps back across the portage trail, it is the beautiful carpets of the moss *Polytrichum juniperinum* that entice me. These pockets of dense green on the forest floor send up brilliant red and yellow stalks with elegant white spires at their tops. Like a futuristic city from a 1950s science fiction novel, the contrasting colours and graceful lines appear otherworldly. Mosses have been captivating the world with their particular style of beauty for a long time. They are ancient plants. The earliest mosses originated about 350 million years ago, or more than 200 million years before the first flowering plants arose. They have maintained a relatively primitive form and have never evolved the necessary structure to produce true leaves, stems or roots. This is why you will never find a tall moss.

Mosses reproduce sexually, but their life cycle is very different from what we would think of as usual. What we typically recognize as a moss-the main body that forms the enchanting green carpets, is actually the stage in the plant's life-cycle where it has only one set of chromosomes. In humans, two sets of chromosomes come together, one in the egg from the mother and one from the sperm of the father, to create a new individual, which has two sets of chromosomes in every cell. We consider ourselves an individual only once the chromosomes of both of our parents come together to make us who we are. But the mosses turn this concept on its head. The two sets of chromosomes only come together briefly to produce the spores that will then give rise to the new individuals, which will only have one set of chromosomes. The difference between this and what occurs with most other sexually reproducing plants and animals is that the large, long-lived stage is the one with only a single set of chromosomes.

So although we think of the sperm and egg as the means to produce a new human, all we base that upon is the fact that we last so much longer and seem much more complex than a sperm or egg. But you could argue that our bodies are just a mechanism that sperm and eggs have come up with so that they can mix together reliably and reproduce themselves. This is how it would appear from the mosses' point of view.

When my two chromosome per cell self imagines a carpeting of moss on the forest floor, my minds eye brings up a picture of deep shades of lush green, where tiny droplets from an early morning dew, or from a lightly fallen rain, glisten in the early morning sun. Perhaps that's just romanticizing, but this idealistic picture is not so far from the truth. Mosses do live in places with a lot of moisture because, unlike flowering plants, whose pollen is usually delivered to the egg by the wind or insects, mosses need at least a few drops of rain or dew in order to become fertilized. This is because, believe it or not, the mosses produce a swimming sperm, a feature which they have retained in common with the aquatic algae.

When I first learned that plants could have swimming sperm, it threw a wrench into all of my unspoken assumptions about how life is split up into neat little categories. Plants once seemed so very different from animals. It always seemed that they just sat there and had things happen to them; they weren't out there actively doing stuff. Of course, as we have seen, the plants are very active in promoting their own survival, but finding out that they have sperm that swim around looking for an egg was a watershed for me. Maybe the neat little compartments that I used to categorize life had more to do with the limits of my mind than they had to do with the way the world actually works.

Below their elegant spires, the green carpets of moss are composed of numerous separate shoots packed closely together. Some of these shoots will hold the male sex organs, which produce sperm, and others will contain the female sex organs, which produce eggs. On a suitably wet day, the sperm of a male shoot will be released and either swim through the thin layer of moisture on the leaves, or be splashed by the rain onto a female shoot. Once on the female shoot, they must swim to the female sex organ and then down its neck until they reach the egg. It is this arduous path that will finally bring two sets of chromosomes together and produce the brilliant spires whose spores will give rise to the next generation of moss.

Along this portage trail, the rain has certainly given the moss a shiny wet covering, but I look in vain for the shaft of sunlight to make it sparkle. The rain has mostly let up and has become such a light drizzle that I need to stop and look carefully to be able to tell that it is raining at all. This trail isn't as wet as the other, but I still need to step around or over the occasional puddle if I don't want to be up to my ankles in water. The two trips across this Class II portage have done their job. I've stretched my legs and am happy to paddle on. The section of creek that this portage just avoided is fed from two directions. From the south, a seemingly motionless ribbon of water meanders through a marshy area filled with the brown dried stalks of last year's cattails. I decide that I will paddle through the marsh tomorrow. Today I will head west, along a straighter route, which arises from a lake only a short distance away.

Back in the Water: Water-Lilies and Water Striders

From the canoe, I can see the fresh new leaves of water-lilies, still a few feet below the water's surface. The leaves appear to be reaching upwards, stretching towards the light, and yet I know they are not. Water-lilies have found a more efficient way of getting their leaves to the surface. It's a simple solution. Their leaves float. Floating leaves can get to the surface even in the murkiest water, with no effort at all on the part of the plant. They also eliminate the need for a supporting stem, saving precious resources for other uses. Having roots anchored several feet underwater, and a leaf which floats above the water, has allowed water lilies to become widespread and abundant plants in our lakes and ponds. These beautiful plants have undergone many useful adaptations to make their way of life possible, and in the process have become a vital food source for numerous animals, from insects to moose.

One of the main challenges that a water lily faces is its need to exchange gases with the outside air. The leaves of most plants have hundreds of thousands of tiny pores on their underside which allow carbon dioxide to diffuse in from the surrounding air. This obviously would not work with a floating leaf. The leaves of water-lilies have had to adapt their structure so that their pores are on the upper side of the leaf, where they have access to the carbon dioxide in the air. As I just mentioned, they have also adapted the structure of their stem. It does not need to support the plant and so is flexible rather than rigid, drifting back and forth with the movement of the water. Their stems have also evolved tubes that allow air to get from the leaves above the water, down to their underwater roots.

All of these adaptations have forced water-lilies to adapt in another way as well. It takes a lot of energy to grow from the bottom of a pond to the surface of the water. Until their leaves have reached the surface where they can breathe, they need to rely on energy stored in their roots from the previous year. So all summer long they accumulate any excess carbon that they don't need for day-to-day survival or reproduction, and store it in a thick fleshy rootstalk to prepare for the following year.

This carbohydrate-rich rootstalk has become an important source of food to those animals that have adapted to take advantage of it. It is not an uncommon sight for canoeists to see a moose standing up to its armpits in the river, munching on these plants and escaping the flies. The rootstalks can also be cooked like a potato if you wish to use the stored energy for yourself, and have in fact been used in this way for many centuries by First Nations people. But the animal that uses the rootstalk to the most advantage is the beaver. In the summer, water–lilies are the most important food source of beavers. They will eat the whole plant, from the rootstalk right up to the flower itself, taking advantage of all of that energy that the plant is attempting to store for itself. Perhaps not coincidentally, water lilies thrive in the very conditions that beavers create when they dam a creek or river to make a pond. Dare we say that beavers are farmers?

When summer comes, this little creek will seem smaller, and more lush than it does now. The water lilies and other aquatic plants will decorate the edges. Their flowers will bloom, inviting a host of insects to pollinate them. Dragonflies will dart overhead, and frogs will sit stereotypically on lily pads, waiting for an insect to come too close. Today, the only insects I can see are two lonely water striders. The rain has let up and the water is calm, leaving them free to climb down off the shoreline vegetation and go for a swim. Water striders overwinter as adults under rocks or logs at the bottom of the lake or pond where they live. These two water striders appear to be wingless. Although most water striders are wingless, occasionally they are born with wings, which they will keep throughout their entire lives. A female will have winged rather than wingless offspring when the conditions around her have changed and no longer seem ideal for the survival of her young. This may be because the pond or puddle she is in is drying out, or because the local water strider population is becoming overcrowded, signaling to the mother that her offspring might have a better chance elsewhere. These water striders can then fly to a new location where they will have wingless offspring and start a new population. If you've ever found water striders in a puddle in the forest, or even in an unopened swimming pool and wondered how they got there, this is their secret.

Dinner Time? Bog Rosemary

Paddling along the shore, I can see that the bog rosemary (*Andromeda glaucophylla*) is in flower. The leaves of bog rosemary look uncannily like the leaves of the true rosemary that you might find in your spice rack. Both are thin and dark green with their edges rolled inward. The two, however, are unrelated. Like leatherleaf and mayflower, bog rosemary is in the Heath family (Ericaceae), a family of plants that is well adapted to surviving cold winters. True rosemary (*Rosmarinus officinalis*) is actually a member of the Mint family (Lamiaceae), a family that also includes oregano, thyme, basil and sage. Like most members of this family, it can be quickly identified by its stems, which are square in cross section.

Being native to the Mediterranean region, true rosemary will not survive over the winter if the temperature drops below about 4°C. So if you are one of those talented canoeists who prepares more extravagant meals in the bush than you do at home, you might want to think twice before substituting bog rosemary for the usual rosemary when cooking up your Rosemary Grilled Chicken with Wild Mushroom Sauce entrée. Bog rosemary leaves contain andromedotoxin and cause low blood pressure, difficulty breathing and intestinal upset. Your use of wild foods would not be appreciated on the night you made this dish.

I paddle into the lake that feeds the creek. It's a pretty little lake with mostly conifers lining the shores. The water is mud-brown in colour, and there is a marshy area feeding the lake from the far side. I don't think the lake can be very deep. While paddling along the shore, I hear the deep bass drumming sound of a ruffed grouse across the lake. Like the call of the loon, this is another one of those sounds of the forests that makes me feel like I am a part of a greater natural world.

Selecting for Volume: The Grouse

To try to attract a mate, the male grouse will stand prominently atop a log, displaying his beautiful tail and beating his wings against the air. The speed and pressure of his beating wings create a vacuum that produces the characteristic deep resonating sound, similar to how lightning creates thunder. Female grouse are attracted to this display because selection pressures have favoured females who choose a robust, healthy mate. By displaying in such a grandiose fashion the male is in essence saying "Look at me, I am such a fantastically fit individual that I can stand out in the open and draw all sorts of attention to myself while still being able to avoid predators-you should want my genes in your children". It is this sort of evolutionary pressure that accounts for the beautiful colours and impractical accessories of many male birds.

After mating, the female grouse will lay about eight or ten eggs in a shallow depression lined with leaves. The chicks are precocial, meaning that as soon as they have dried out after hatching they are ready to leave the nest and start feeding themselves. Although only the size of a human thumb, they can fly after only five days. It's a treat if you can ever spot one of these little birds zooming by like a giant bumblebee.

I stop at the first campsite I come to and leave my canoe perched on the shore while I go up to have a look. It's a nice campsite nestled beneath tall red pines. The understory is made up of white pine and balsam fir, and there are a few tall dead birches still standing amongst the large pines. This mix of trees indicates that succession is taking place. The old birches have already died, and when the big red pines eventually die, they will be replaced by the white pines and balsam firs that are now still young. Both red pines and birches are intolerant of shade and do not grow with other trees blocking the sun above them. The white pines are reasonably tolerant of shade, however, and balsam fir is very tolerant of shade when they are young, becoming less so as they age. As a result, they will grow to replace the now towering red pines and the already dead birches. For tonight, however, the stately red pines make for a beautiful campsite. Below, wild strawberry (*Fragaria virginiana*) is in flower, and the delicate leaves of early meadow rue (*Thalictrum dioicum*) rise from the forest floor. I decide I will call this home for the night.

It has stopped raining, so I carry my gear up from the canoe and get to work on setting up the tent before the rain starts again. From the numerous pinecones I have to move, I suspect that no one has used this site yet this year. The air has become perfectly still, the way that it sometimes does between rains on an overcast day. I take a walk along the trail to have a look at my washroom facilities for the evening. Along the trail, the leaves of *Epigaea repens* (mayflower) are abundant, but the flowers have not yet bloomed. The distinctive rounded leaves of *Hepatica americana* (round-lobed hepatica) also dot the trail. At the end of the trail, the Park has set up their usual wooden box toilet seat. Most sites have this simpler style box as opposed to the enclosed outhouse style po-po. When the bugs aren't bad, you can't beat the box. No matter how much marble or how many jacuzzis you fill your washroom with, it can not compare to the beauty of a box seat in the open air of the wilderness.

Before dinner, I paddle to a portage on the far side of the lake to have a look around. The portage trail leads through a short section of marsh before reaching the solid ground of the forest. There is a small wooden dock followed by a short stretch of cut logs leading through the marsh and making it easier to walk through the muck underfoot. The logs look as though they've been around for years, and with the water levels high, they are only partly successful in keeping my feet dry and clean. It is immediately apparent that the forest this portage passes through is different from the others. The understory is so thick with young balsam firs that I can't see into the forest at all. The first trees of the portage are striking. They are two giants, a white pine, and a larch. The white pine, of course, still has all of its needles from last year, but the larch is just beginning to grow this season's new needles. Unlike other conifers, although they have needles, larches shed them each winter.

Which to Choose? Needles VS Leaves

For a tree that lives in the north, where there is only a short growing season, the advantage of having leaves all year is that it extends the length of the growing season. In the fall, the tree can continue to photosynthesize until it has frozen, and in the spring its leaves are ready long before the deciduous trees can risk growing their new leaves. A tree that keeps the same leaves for more than a year also gains the advantage of being able to use the resources that would have gone into growing new leaves for other purposes. This is especially important for trees that have only a brief season to obtain these resources, or that live in environments that have a limited supply of nutrients.

A tree that keeps its leaves through the winter, of course, faces challenges. The major difficulty is that leaves are where most of the gas exchange of the plant occurs. The leaves of a plant have tiny openings, called stomata, which let in the carbon dioxide that allows plants to grow, and release oxygen, the waste product of photosynthesis. Through these same openings in the leaf, water also evaporates into the air. In the summer, the evaporation of water through the stomata is a good thing and is what draws the water up from the roots to the top of the plant through a process called transpiration. But when the water in the ground is frozen, the plant can't draw fresh water from the ground and so any loss of water through the leaves will risk drying the plant out. This is why evergreen trees have needles. Needles are, for the most part, an adaptation for allowing a tree to withstand periods of limited water. Apart from having to withstand freezing, the long winter months when the water is frozen mean that trees have to be able to survive with little or no water. This is the main reason why deciduous trees lose their leaves. In the winter, when they are in their leafless condition, the deciduous trees in Ontario have only 4% of their usual water loss. Trees that keep their leaves need to be able to drastically limit their water loss as well, which is why they have adapted their leaves to be less “leaf-like” and more “needle-like”.

Needles have many adaptations that allow them to conserve water. Not only do they have a very small surface area, but they are also surrounded by a thick, wax-coated waterproof covering, often with small hairs so they do not lose water as quickly as a plant with a regular leaf would. As well, their stomata are inside grooves in the needles, so that the water does not evaporate as quickly.

Why the larches have adapted to have needles they lose each year is still a bit of an open question. In the far northern regions, where even the conifers have trouble with snow damage and water loss, losing their needles does seem to give them an advantage. This may largely be because in the north, the soil often remains frozen very late into the spring, long after the air has heated up enough to cause needles to lose water. Dropping their needles would allow them to avoid excessive water loss at this time of year. There is one thing that is clear, however, and that is the success of this strategy. The larches have managed to combine the advantages of both the coniferous and deciduous trees, which has allowed them to grow further north and in harsher conditions than almost any other species.

Beyond the giant larch tree, the portage trail runs along the side of the marsh, seeming a lot more northern and undisturbed than the forests I've seen so far today. The trees along the path are mostly giant old pines, with a few old dying birches still standing tall. The understory of the forest is crowded, not only with young balsam firs, but also with fallen trees and dead branches, all in different stages of decomposition. Although just across a small lake from my campsite, the lay of the land and the fact that the forest here faces north and therefore doesn't receive very much sunlight, results in a forest consisting of more northern species. These towering white pines, themselves somewhat tolerant to shade, will likely be replaced by the shade-tolerant balsam firs, which will form another kind of climax forest, better suited to the more northern microclimate along the southern edge of the marsh. This is one of the beautiful things about this part of the country. This northern forest with huge pines and larches gradually being replaced by balsam firs is just a portage away from a forest typical of more southern areas, with oaks, maples and beeches.

The floor of this forest is covered not with leaves, but with a thick layer of pine needles. The understory plants growing here are also very different from the plants I've seen so far. Ferns, and the splotchy, beat-up leaves of last year's bunchberry plants (*Cornus canadensis*) are scattered around the deadfall, replacing the trailing arbutus and meadow rue of the previous portages. Club mosses (*Lycopodium obscurum*), tiny plants that at first glance look like they might be pine seedlings, make up small patches of deep green throughout the forest. I can remember when I was young, I would pick these little plants to use as pine trees on my model train set. I didn't realize at the time that in some areas they were becoming endangered because so many people were collecting them, usually to use as Christmas decorations. But here, and in the forests away from towns and cities, they are abundant.

Defining our Civilization: Club Moss

Club moss is actually a misnomer; they are not actually mosses at all. Unlike the mosses that we talked about earlier, club mosses have true roots, stems, and leaves. Also unlike the mosses, the little green club mosses that I can see on the forest floor have two sets of chromosomes in each cell. Like most other plants and animals, the stage in a club mosses life-cycle where they have only one set of chromosomes per cell is the more unobtrusive and less enduring one.

All species of club mosses are in a division of plants called the lycophytes, a widespread but inconspicuous group that is found all over the world. The lycophytes weren't always so inconspicuous, however. Three or four hundred million years ago, before the age of the dinosaurs, and long before there were any flowering plants, lycophytes dominated the landscape. Some were treelike plants over 50 meters tall and twenty feet around their "trunks". Surprisingly, these giant lycophytes that lived so long ago have a huge impact on each of our daily lives. Based on the decisions we make now, lycophytes may in fact determine our security, peace and survival in the future. This is because the remains of these giant lycophytes that once dominated the landscape are now underground, making up most of the earth's coal and oil reserves. Our entire economy, and the foreign policy of many nations revolve around the coal and oil these ancient lycophytes have become. As we start wars and become increasingly obsessed over control of these reserves, the lycophytes may have the last laugh. We are releasing literally millions of years of these plants' accumulated carbon over just a short couple of hundred years. And just as a change in climate ended their dominance, the much more drastic change in climate that we are now causing will very likely end the reign of today's dominant life forms. Long after we are gone, who knows what will thrive in the new conditions we created. Perhaps the lycophytes will rise again.

As I walk along the trail, the logs, branches and entire trees that lay scattered haphazardly on the forest floor, give the trail an earthy feel. I can still hear the drumming of the grouse. With all the fallen trees lying around, there are plenty of platforms for a male grouse to display from. The deadfall also provides many places to hide. It's a funny name, deadfall. A big, healthy tree in the prime of its life is actually made up almost entirely of dead wood. Only a small portion around the outside of the trunk is live tissue. But once a tree becomes deadfall and begins to decompose, it is absolutely teeming with life. First, the fungi and bacteria invade, replacing the dead wood with their own living cells. Insects move in, and before long mammals, birds and reptiles will be taking shelter or looking for a meal in the rotting wood.

The Decomposing Forest

The forest is in fact full of life that we can't see. In the dead trees and in the soil, even in the leaves and roots of living trees, there are literally tons of fungi and bacteria every direction that we look. The plants, which are the life form that appears to dominate the forest, are of course of paramount importance. Because plants are virtually the only organisms that can harness the power of the sun to produce life (apart from a tiny number of specialized protists and bacteria), they are the absolutely essential first requirement to life on Earth. By mastering the trick of photosynthesis, they produce all the oxygen that we breathe and are the ultimate suppliers of all the food we eat. Even when a wolf eats a deer, it is feeding on the energy of the sun, captured in plants and then turned into deer meat when the deer ate the plants. But even though the plants provide the raw material and energy required for life to function, it is the microorganisms, the fungi and bacteria, which allow the complex ecosystems that we have today to exist.

Perhaps the most fundamental role of the microorganisms is their role as decomposers. The decomposers feed on the dead. By breaking down the remains and wastes of other organisms into simple products such as nitrates, phosphates, and carbon dioxide, they ensure that these essential molecules are available to be reused by the next generation of plants. As decomposers, the fungi are unparalleled. They have evolved the ability to break down the cellulose and lignin that wood is made of. Without this ability, the trunks of trees would pile up on the ground until there were no longer any nutrients left in the soil. The lignin and cellulose of dead trees are carbon sources that are unavailable to almost all other organisms, making it the fungi that drive the global carbon cycle by releasing the CO_2 trapped in the dead tree and making it available for new plant growth.

Walking along a wooded trail, the plants seem to dominate the landscape, leaving the casual observer with the impression that there are far fewer fungi than plants. But the fact is that the fungi are just more inconspicuous than the plants. In a typical forest, the only obvious sign of fungi is the odd mushroom poking up through the leaves on the forest floor. Although often beautiful, and impressive in their diversity, their scattered appearance might give you the feeling that they play only a minor role in the forest. The mushroom you see, however, is just the tip of the iceberg. Mushrooms are the reproductive structures of some fungi, much like the flowers of plants. Imagine if all that you could see of plants was their flowers. Most of the year, you would see nothing at all.

The mushrooms have networks of microscopic hyphae in the soil, which make up the main body of the fungus. Hyphae act much like the roots of plants, spreading out through the soil or wood and collecting nutrients for the individual fungus. Unlike plant roots, however, the hyphae also collect carbon and are typically extremely fine, delicate structures that form a kind of web through whatever they are growing in. What starts out as a spore, a microscopic point on the forest floor, can grow through the soil until a single individual can have hyphae that are spread out over an area hectares in extent. In fact, the largest fungal individual ever found is from the species *Armillaria gallica* or honey mushroom. A common species in our forest, this one individual is at least 2500 years old, and its enormous underground body covers an area of 70 hectares (173 acres) in extent. It is estimated to weigh about 400,000 kg. For comparison, a blue whale weighs about 100,000 kg. My Master's research focused on trying to work out the mutation rate within this individual. Individuals like this one can shuttle nutrients from one section of their huge underground body to another, allowing them to expand through inhospitable sections of the forest. All you would see of such an individual is the odd mushroom scattered throughout its huge territory; everything else is either underground or within decomposing wood on the forest floor.

Each of the tremendous diversity of mushrooms in the forest has an underground body of its own. Not all species have individuals that grow as large as *Armillaria gallica* does, but they all have more going on below the surface than above it. With some species, you can see approximately how big an individual is. Fairy ring mushrooms (*Marasmius oreades*) tend to produce their mushrooms at the outer edges of their underground bodies after it rains. As a result, the mushrooms form a ring that outlines their shape, at least in two dimensions. Each year, as they grow, the fairy ring of mushrooms will become larger. Other species grow exclusively on decomposing wood. These individuals might be confined in size to the piece of wood they are growing in.

A fungal individual with its hyphae growing through a dead tree may not be able to expand beyond the size of the tree that it is living in (if it can grow even this large, it does have to compete with any other fungi, either of its own species or another, growing in the same tree). Its future lies in the mushrooms, or other spore producing bodies you see on the surface of the wood, which will send their spores out in the hopes that the next generation will find a suitable spot to grow. But whatever the size of each individual fungus, the soil and wood are full of living fungal tissue that we can't see.

Fungi: Far More Than Just Decomposers

The reach of the fungi extends far further than to just those that produce mushrooms and decompose wood, however. Most species of fungi are even more inconspicuous, and don't produce easily recognized reproductive structures at all. As well, not all fungi are decomposers. A large number of fungi play roles that are just as important as the decomposers, even though they can't break down plants to obtain carbon for themselves. They instead have above ground carbon collectors. We call these plants. There are thousands of species of fungi that have evolved the ability to obtain carbon from living plants. Perhaps the most important of these, and to me the most interesting, form what are known as mycorrhizal associations with the roots of plants. Some of these fungi are of the mushroom-producing sort and some are not, but many of the mushrooms that you see as you walk through the forest are from fungi which have this mycorrhizal relationship.

The fine networks of hyphae of these mycorrhizal fungi will penetrate the roots of plants and form specialized structures inside the cells of the root. The idealized pictures of root structure from your grade ten biology textbook rarely exist in nature. Most roots are full of structures that do not belong to them at all, and many are covered with layers of hyphae on the outside as well. The fungus supplies the plant with nutrients, particularly phosphorous, and the plant supplies the fungus with carbon. So instead of being decomposers and obtaining their carbon by breaking down soil or wood, these fungi get their carbon directly from living plants. This is another example of a symbiotic relationship, where evolution has selected for species that have cooperated with other species to their mutual benefit. This symbiotic relationship likely started 400 million years ago and allowed plants to colonize the land. In fact, 85% of all plant species alive today actually require their fungal partners to survive. This includes many trees, including all of Canada's beautiful conifers.

These plants have evolved together with their fungal partners for so long that they can no longer live without each other. This cooperative relationship, although beneficial to both the plants and fungi involved, appears to be dominated by the fungi. Plants breathe carbon dioxide, so one might expect that as the burning of fossil fuels causes the atmosphere to have a higher concentration of carbon dioxide, some of this would be taken up and stored, making trees bigger and bushier. This doesn't tend to happen. Instead, their root masses and their mycorrhizal partners increase in size. The fungi get the carbon. The fungi may have even more control than this. Often the hyphae of one of these fungal individuals have this mycorrhizal relationship with many plants at once, even with plants of different species. These fungi can shuttle nutrients, including carbon, between these plants however suits them best. The fungus can maximize its own growth by starving some trees and feeding others. Should we perhaps think about most plants as just big carbon collectors for fungi?

Other fungi, just as ubiquitous as the mycorrhizal fungi, have co-evolved in a symbiotic relationship with the leaves, rather than the roots of plants. These fungi are called endophytes. Endophytic fungi are even more common than mycorrhizal fungi and inhabit the leaves or needles of almost all plant species, from the tallest trees to the smallest blade of grass. Endophytic fungi are even found in the tissues of algae. The fungus receives nutrition and protection from the host plant while protecting the plant from herbivores and pathogens.

There are many different species of fungi that live as endophytes, some of which seem to be generalists and some of which are specialists. The generalists give broad-based protection against insect herbivores by utilizing the fungi's ability to produce toxic or unpleasant compounds. The specialists direct their attack towards specific organisms that are pests of their host plant. Some species produce compounds to reduce needle attack by specific insect borers. Others deter the larvae of certain insects from feeding on leaves. Others even give their host plant increased resistance to other fungi which are specific pathogens of the plant. One species colonizes aphids that feed on the leaves of their host plant. They eventually kill the aphid, but not before hitching themselves a ride to another leaf, which they can then colonize. This means that this fungus has more than one niche that it is adept at surviving in. It can thrive in a cooperative relationship with a plant, and also as a pathogen of an insect. These are two very different environments, and these fungi can exploit them both, while at the same time using the insect as their method of dispersal. This flexibility of the fungal endophyte has been found in many other species as well.

The importance of endophytes in the ecosystem seems to continue even after the leaf that they reside in has died. Many of the species of fungi that are found as endophytes in healthy leaves are also associated with the early stages of litter decomposition of these same plants. So, when the leaf dies, the endophyte can change its role once again and become a decomposer. As more research is done, I expect that the role of endophytes in ecosystem function and plant community diversity will become recognized as even more significant than it is already known to be.

More Sperm: The Ferns

The portage trail isn't long (155m), and I don't need to carry anything, so I can appreciate the deadfall, rather than curse it for blocking the trail. The trail is damp, and the end is nearly as swampy as the beginning. In the marshy field at the end of the trail, I can see dozens of stout brown stalks standing resolutely above last year's now brownish greenery. These are fertile blades produced by last year's sensitive ferns (*Onoclea sensibilis*). Like fungi and mosses, ferns produce spores rather than seeds. Spores are typically much smaller than seeds and contain very few stored food resources. With seeds, fertilization has taken place before the seed leaves the parent plant and the genes of its two parents have already mixed and produced an embryo. Spores, on the other hand, usually only contain the genetic material from one parent, and fertilization takes place only after the spore leaves its parent and combines with the genetic material from another spore.

Most ferns have their spores clustered into rows of small dots, called sori, on the back of their leaves. There are a few ferns, however, including the sensitive fern, whose spores are on special stalks arising separately from the base of the fern. These fertile stalks have an advantage over sori in that they are able to persist through the winter and release their spores in the early spring. This gives them an early start to the season and allows the wind to successfully distribute them before their leaves emerge and other plants begin to crowd around them.

The spores of these fertile stalks, like the spores on the back of the leaves in other species, are not actually sex cells, but are produced asexually. Hence, both the leaves and the fertile stalks of the fern (which make up the entire plant that we typically recognize as a fern) are called the sporophyte generation, and are only one of two stages in the life-cycle of ferns. The spores that are released from this generation, if they land in a suitable location, will germinate but will not produce another sporophyte, with its fern-like leaves and fertile stalks. They will instead produce what is called a prothalli. The prothalli is a small leaf-like structure, usually less than a centimeter in diameter, that is often heart-shaped. They can be found growing just above the moss and debris on the forest floor. It is entirely independent of the sporophyte from which it came, and must obtain the carbon it needs through photosynthesis just like other plants.

The prothalli is short-lived and, unlike the sporophyte generation, does not have true roots, but is anchored to the soil with tiny hair-like structures called rhizoids, which serve as roots. Although insignificant looking in comparison with the sporophyte generation, it is the prothalli, not the plant we typically recognize as the fern, that will form the sex organs and produce the eggs and sperm. Like the mosses, ferns are another plant that has swimming sperm, and these are released by the prothalli and will swim until they find an egg to fertilize. In some species, the egg may be on the same prothallus, while in others the prothalli are each separate sexes, and the sperm will have to find an egg on a neighboring one. When the sperm and egg do fuse, the resulting cell will develop into an embryo and grow into the sporophyte generation, or what we recognize as the fern again.

Building Complex Relationships: Bog Plants

I return to the canoe, walking back over the partly submerged muddy logs at the start of the portage and push my canoe out into the flat water of the lake. The air has become still, and the overcast sky has broken apart in places, displaying the bright blue behind. Rather than paddle across the lake directly, I instead paddle slowly along the marsh on the western shore. The marsh is thick with the brown leaves of sedges and grasses. Leatherleaf (*Chamaedaphne calyculata*) is scattered throughout the marsh, its flowers hanging charmingly along the bottom of its branches. The plant that most catches my eye, however, is the cranberry (*Vaccinium macrocarpon*). It catches my eye not because the plant is particularly conspicuous, but because hundreds of bright red berries still adorn them. The name cranberries came about when German and Dutch settlers called these plants crane berries because their blossoms look like the head and bill of a crane. This is another type of complex flower that has adapted to be pollinated primarily by bumblebees. A single bumblebee can visit 50 to 100 cranberry blossoms in one collection trip. Bumblebees are so efficient at gathering pollen from these flowers because they have developed a technique known as buzz pollination. The bee will cling to the flower with its feet and vibrate its wings at twice the speed it normally would to fly. The flower has adapted to respond to this with a puff of pollen that the bee can then store in the pollen sacs on its legs.

Another plant that is predominant along the shore of the bog is sweet gale (*Myrica gale*). Unlike many of the plants we've seen, sweet gale loses its leaves in the fall, and the numerous plants in this bog have not yet begun to grow their leaves for the new season. The flowers, on the other hand, sit along the top of each stem, making the bare twigs that arise from the water easily identifiable.

The flowers of sweet gale themselves look more like leaves, or perhaps some sort of cone, but a closer examination reveals a unique beauty with subtle hues of browns, purples and greens. Like many of the plants that grow in boggy areas, sweet gale has difficulty acquiring nitrogen. Because they are always wet, the nitrogen in the soils of boggy areas is continually being leached away, and so the plants in bogs must find ways to obtain nitrogen other than from the soil.

Sweet gale has overcome this lack of nitrogen in the soil by forming a symbiotic relationship with a species of bacteria from the genus *Frankia*. In yet another example of co-operation between species, these plants create nodules in their roots specifically to give *Frankia* a place to live, where they also supply the bacteria with some of the carbohydrates they produce through photosynthesis. They essentially give the bacteria free room and board. They do this because *Frankia* is one of a small number of species of bacteria that can convert nitrogen in the air into ammonia. This ammonia can then be converted into nitrites and nitrates, which are the forms of nitrogen that plants can use. This process of converting atmospheric nitrogen into nitrites and nitrates is called nitrogen fixation, and no organism more complex than bacteria can accomplish it. So in exchange for food and shelter, *Frankia* produces a supply of nitrogen for its host.

Most people are familiar with the fact that the legumes, peas and beans, can be planted in order to enrich the soil with nitrogen. Like *Myrica gale*, this is also because of a symbiotic relationship with nitrogen-fixing bacteria, in this case from the genus *Rhizobium*. In fact, nitrogen-fixing bacteria in symbiotic relationships with plants account for a full 50% of the atmospheric nitrogen that is fixed on Earth.

In order to work together to fix nitrogen efficiently, there is a hurdle these symbionts need to overcome. Because of the chemical pathways involved, *Frankia* and *Rhizobium* can only fix nitrogen where there is no oxygen. The bacterial symbiont, *Frankia*, limits the amount of oxygen in its vicinity by producing specialized vesicles with as many as fifty layers of lipids. But the plants have also undergone an interesting adaptation to remove oxygen from their root nodules and protect their bacterial symbionts. Their root nodules produce a form of hemoglobin, similar to the hemoglobin that carries oxygen through our blood. This hemoglobin (leghemoglobin as it is called in the legumes) binds any free oxygen in the nodules that would inhibit the process of nitrogen fixation, leaving the bacteria free to produce nitrogen for the plant. Because it is the hemoglobin that gives blood its red colour, if you cut open one of these tiny nodules, it will be a subtle blood red.

I paddle into an inlet in the marsh, where the cranberries, protected from the wind, are even more numerous. Scattered along the edges of the water on either side of me are the impressive purple and green leaves of pitcher plants (*Sarracenia purpurea*). A rosette of elegant upright tubes, the leaves of the pitcher plant singles it out immediately. The early spring rains mean that most of the tubes are half filled with water. Those right along the edge of the water are completely submerged, the lake still swollen from the melting snow. I love these plants with their elaborate tubular leaves and unusual, almost prehistoric-looking flowers. I, in fact, had a cat named Sarracenia.

Like sweet gale, pitcher plants have had to adapt to a low-nitrogen environment as well, but they have evolved a much different strategy. Instead of having their nitrogen created for them, they use nitrogen that is already in the ecosystem. Their beautiful tubular leaves attract insects. Once inside, the insects find that the tubes are designed to keep them there. A combination of smooth walls and recurved hairs lining the inside of the tube means that most insects have difficulty crawling out and eventually fall to the bottom of the tube and drown. Then, as the insects decompose, aided by enzymes the plant secretes into the water, the pitcher plant absorbs the nitrogen as well as other nutrients that make up the body of the insect.

Not every species of insect that enters the tube are simply prey, however. There is often an entire miniature ecosystem inside each of these tubular leaves. Living in the fluid at the bottom of these leaves, there will be a variety of bacteria, protozoa, and aquatic mites. They may also house the maggot of one of several species of flesh fly, the larvae of the midge *Metriocnemus knabi* and perhaps the larvae of the mosquito *Wyeomyia smithii*, which is a species of mosquito that breeds exclusively in pitcher plants. These organisms don't only survive inside the hostile environment of the Pitcher Plant leaf, but in fact have adapted to thrive there and build a relationship with their host.

Although the plant can release digestive enzymes of its own, the bacteria, protozoa and insects that have adapted to living inside the leaves are far more efficient at breaking down the bodies of the hapless insects that fall into the pitcher plants' trap. Each of these creatures living inside the leaf has its own niche. Flesh fly maggots feed mainly on fresh carcasses at the surface of the water, whereas the larvae of the midge feed on the slightly more broken down carcasses at the base of the leaf. By cutting up and boring holes into the bodies of the trapped insects, they expose the insects' insides, and make it much quicker for them to be broken down by the bacteria and protozoans living there. As they eat, the bacteria and protozoans will multiply in number and be preyed upon by the mosquito larvae.

The waste products produced by each of the pitcher plant's guests are a variety of nutrients, including the nitrogen that is in such short supply in bogs and marshes. But because the organisms living in the pitcher plant breathe oxygen, there is another waste product that they produce, which is of course carbon dioxide. So, as well as getting help in breaking down their prey to obtain the much needed nitrogen, pitcher plants also get carbon dioxide from the insects in their pitchers. This miniature ecosystem works in the same way that our larger, more familiar ecosystems work. In the larger ecosystem, humans and other non-photosynthetic organisms breathe oxygen created by plants and expel carbon dioxide, which plants "breathe". Inside the pitcher plant the tiny organisms are releasing carbon dioxide as they breathe, and this carbon dioxide is taken up by the pitcher plant. The pitcher plant, for its part, infuses oxygen (the waste product of photosynthesis) into the water in its pitchers, allowing its tiny ecosystem to flourish. This flourishing ecosystem will then produce more carbon and nitrogen for the pitcher plant to use.

This system is complementary in another way as well. It is on hot, sunny days when the air is still that the limiting factor for pitcher plant growth is the amount of carbon dioxide in the air. This is because the three most basic things that a plant needs for growth and reproduction are water, sunlight, and carbon dioxide. The energy from sunlight converts the water and carbon dioxide into carbohydrates, which supply the plant with both the energy and carbon needed to create all the plant's molecules. In a marsh or bog, there is plenty of water, and on a hot sunny day it is the availability of carbon dioxide that will be the limiting factor for the plant. But, as anyone who has left garbage out in the hot sun knows, it is precisely on these hot still days that the decomposition going on inside of the pitcher will be at its quickest. Heated by the sun, the inhabitants of the pitcher plants will produce more carbon dioxide and require more oxygen than on a cool day. So, the inhabitants of the pitcher plant produce the most carbon dioxide at the time when the plant needs the most carbon dioxide.

A Canoeists Favourite: Mosquitoes

As one of the inhabitants of the pitcher plant, I think a quick word on mosquitoes is in order, seeing as they are so close to most canoeist's hearts. Many people, and justifiably so, will plan their canoe trips to avoid the worst of mosquito season. Everyone knows that the places we canoe are busiest in July and August, and although the weather may be beautiful in June, most people would rather battle the human crowds than the legions of mosquitoes. Even when you don't design your entire trip around them, they will often mark almost every decision throughout your trip. When choosing a campsite, we will select a windy point well away from marshy areas. Once our camp is set up, we jump in and out of the tent, quickly zipping it behind us as though pursued by demons. We build smoky fires to fend them off and may even go to bed three hours before we are tired just to get away from them. On the portage trail, we wear long pants and shirts on the hottest, most humid of days, and will even resort to smearing poison all over our faces and hands. So, it is perhaps a gross understatement to say that mosquitoes have a relationship with canoeists.

Whether canoeists are of such importance to mosquitoes is of course a different question. It may not be news to canoeists that there are a lot of mosquitoes in Canada's forests, but the large number of different species of mosquito is perhaps surprising. Worldwide there are approximately 3500 different species of mosquito. Algonquin Park alone is home to 27 different species. Each of these species has a different niche they have adapted to survive in, and their presence throughout the year is the result of a succession of different species, each appearing at a different time. Most species live for less than a month as adults and although about 80% of the mosquito species living in Canada take blood, there are only a small number of these that are abundant enough to be considered pests.

The first mosquitoes to appear in the spring are one of several species whose already fertilized adult females have overwintered by the thousands in the rotten wood of a hollow tree. These species from the genera *Anopheles* and *Culiseta* will be the first to be seen looking for a meal of blood. Most species, however, overwinter as eggs and must develop through their larval stages in small pools of meltwater, or in the still water of a marsh or pond. Of these, there are several species that account for the tremendous surge in the number of mosquitoes in the spring. These are mosquitoes from the genus Aedes, and in the fall they will lay their eggs in a depression in the ground, or any other place they think is likely to fill with water the next spring. When the snow melts, these eggs hatch and go through their larval stage, usually emerging as adults en masse about the time the leaves are opening on the trees. Like the trees, these eggs will take cues from the environment to determine when to hatch. Some wait for the water to reach a certain temperature, while some hatch in response to the byproducts of microorganisms in the water, indicating to them that there will be food for them when they become larvae. Often their mother will die in the same depression that she has laid her eggs in, and the young will benefit by eating the microorganisms generated from her decomposing body. These temporary pools provide an ideal place for their larvae to grow. By developing quickly and in such large numbers, any potential predators are faced with the problem of a vast supply of food for a few days that abruptly disappears. So although a large number of predators have adapted to take advantage of this abundant source of food, they are unable to appreciably reduce the number of larvae that survive. As a result, the woods are full of millions of them, and it is the members of this genus, Aedes, that a canoeist will usually find probing their bodies for blood.

I arrive back at my site, happy at the absence of mosquitoes and anxious to prepare dinner. Spending time outdoors always builds an appetite. For me, it's always an appetite for pasta. I dig out my pots and stove and soon have spaghetti noodles bubbling away. I prefer to cook using a camp stove rather than an open fire, partly because I feel responsible for the CO_2 emissions I cause, and I can't always justify to myself the significantly higher quantity of emissions a fire will produce. In a popular park like Algonquin, there is even more reason to limit the use of fire. When I look around this site, there is almost no dead wood in the vicinity. Contrast this with the amount of dead wood where there are no campsites, and it's clear that campfires are having a large impact. How is the next generation of trees going to obtain nutrients? Campfires can be a lot of fun, and I do occasionally have one, but just as with decisions in day-to-day life, I think everyone should consider how much damage they are willing to cause simply for their own pleasure.

It has begun raining again, and by the time I start to eat, it's raining hard. A tarp would have been a good idea, but I didn't bother to bring one. After about half an hour, the rain stops, leaving the lake covered with a light mist, and the sun poking out from behind dark clouds. In the calm, the lake has become quiet. The only sound is the distant chirping of frogs. These frogs, like the plants, would have just lived through being frozen all winter.

The Frozen Frogs

To consider that frogs must survive freezing over winter somehow seems more impressive to me than when plants do the same thing. I guess it's just because a frog has arms and legs, a head and organs, which makes it easier to imagine what it would be like for us to freeze for months at a time. For them to survive, they must undergo changes that radically change the distribution of water in their bodies. Burrowing themselves just below the ground in the fall, they will initiate these changes only after their bodies actually begin to freeze. When a frog does start to freeze, it will move water out of its cells and into the spaces around its cells, while at the same time pumping its cells full of glucose, which acts as an antifreeze.

The deflated and glucose-laden cells of the frog won't freeze at all, not even at very low temperatures, while the extracellular water (the water between its cells) must freeze slowly to give the cells time to make these adjustments. However, for the water to freeze slowly, it is important that it begins to freeze at a relatively high temperature. If it only began to freeze when the temperature was already very low, it would freeze quickly, causing damage to tissues and not allowing the frog time to adjust to the stresses freezing imposes.

In order to induce freezing when the temperature is just below zero, the water in their extracellular spaces contains what are called ice-nucleating agents. In order for ice crystals to form in pure water, enough water molecules need to come together to create a kind of embryo upon which the ice crystal can grow. The tendency is however, for water molecules to disperse rather than grow, and so the water does not begin to freeze until it is well below zero degrees. The other way that water can begin to freeze is if something other than water forms the template around which the ice crystals form. These are the ice-nucleating agents, and they allow the water to begin to freeze at a higher temperature than it would otherwise. Ice-nucleating agents may be proteins or inorganic compounds, but the most effective ones seem to be certain bacteria and fungi. These microorganisms have been found in both frogs and insects. The animal allows the bacteria or fungi to live inside of them in exchange for their ability to initiate freezing at just below zero degrees. Within a day or two these extracellular spaces, which at this point will contain up to 70% of the frogs water, will be completely frozen, while the shriveled glucose filled cells resist freezing, allowing the frog to stay this way until the spring thaw.

I sit quietly listening to the distant chirping of the recently defrosted frogs for a while before rousing myself to get my camp ready for the night. It will start to get dark before too long, and I would like to get my food hung at a leisurely pace.

If there is one rule everybody knows, it's that you should hang your food up at night so as not to attract bears to your tent. In a much-traveled park like Algonquin, this is especially important because the bears are accustomed to people, and the smell of your equipment, tent and even yourself doesn't intimidate them as it normally would. Further north, where the bears don't typically encounter people, it is far less likely that they will bother you. This is fortunate because as you get further north, the trees get shorter, and at a certain point it is impossible to hang your food at all. Happily, the northern range of raccoons doesn't extend this far. Raccoons seem to have no reservations at all about taking your food. If you leave a pack of food outside your tent at night in Algonquin, you may risk attracting bears, but you will almost certainly have your pack chewed through by raccoons.

I don't want anyone chewing through my pack, so I grab it and wander around about 100 metres from my tent to look for an appropriate tree to hang it from. Through the needles on the forest floor, scattered plants poke their leaves towards the sky. Not much seems to be in flower yet. I see more mayflower (*Epigaea repens*), as well as the leaves of meadow rue (*Thalictrum dioicum*). I also find twinflower (*Linnaea borealis*) creeping across some moss. Not yet in flower, this plant is conspicuous only because of its light green leaves contrasting with the brown of the forest floor. When summer comes, it will send branches up four or five inches above the ground, each with two pinkish flowers hanging like little bells. This was the favorite plant of Carolus Linnaeus, the Swedish botanist who developed the method that we now use to classify all living things using two Latin names. Because of his love for this plant, it was named after him.

When a Tree Falls in the Forest: More than Just a Sound

Walking randomly around the forest, it's hard not to notice that the forest floor is a very uneven surface. Some of the unevenness is due to the contours of the rocks below, but much of it is not. Much of the unevenness is produced when a tree blows down in the forest, either because it has died, or because of unusual weather. When this occurs, it will typically rip much of its root system out of the ground when it falls. We have all seen a tree lying on its side with its roots sticking up into the air, sometimes almost forming a cave. When this happens, a whole new environment is created where the tree once stood. The immediate change, if the tree was a significant one, is that there will suddenly be sunshine reaching the forest floor where there previously wasn't. A whole host of plants will begin to compete for this sunlight. Species of plants that might not normally be able to survive in the shade of the trees will have a brief opportunity to grow, flower and produce seeds, which will then be sent off to find another shaft of sunlight where they can do the same. Unless a new tree manages to take advantage of the sun quickly and fill the space, the crowns of the trees above will close the gap, blocking the sun from the forest floor once again.

The fallen tree also changes the environment by becoming a food source. All the carbon and other nutrients that have been stored up over the years inside of the tree are suddenly available, and before long the fallen tree is teeming with life. Some trees, such as cedars, have evolved to take advantage of their own stored resources and the brief opening in the forest canopy when they fall over. A fallen cedar will send up shoots along its trunk, allowing it to be the first to take advantage of its own stored resources. Genetically, these new trees are the same individual as the fallen trees. This individual has just transformed itself from one big tree into several small trees. You can sometimes see four or five cedar trees in an otherwise inexplicably neat row long after the original fallen trunk has rotted away.

A newly fallen tree changes the microenvironment of the forest floor as well. When the roots are ripped from the ground, the topsoil will come with them, forming a wall of earth interlaced with roots at the base of the fallen tree. This earth will dry out and fall in a pile around the upright roots of the fallen tree. As the roots themselves are broken down, this pile of nutrient-rich soil will become even larger. What was previously spread out over a large area is now in a single mound. This mound provides an opportunity for different organisms to take advantage of the new environment. This is not only a rich source of carbon and nutrients, it is also a high point on the forest floor where a plant can get a head start on those around it for access to the sun. Rain running off the tiny hill may also make it drier than the surrounding forest, allowing plants and animals to live in a forest that they might not have otherwise been able to.

The area around the mound will have changed as well. The soil that was ripped away will leave pits in the forest floor, sometimes right down to bare rock. Water will accumulate in these pits, making yet another environment to be exploited, perhaps by salamanders and frogs or by mosquitoes looking for a place to breed. The pits may gradually fill with leaves and branches, blown into the protected holes by the wind or carried there by the melting snow. The stark contrast between the pits and mounds will be softened, and a moist, nutrient-rich depression may be created beside a mound that was once the roots of a fallen tree. It is these remnants of fallen trees that have created much of the uneven ground that I am walking over in the search for a place to hang my food.

I find a beautiful red pine along the water's edge with a solitary branch, stout enough to hold my pack. I've been carrying a small rock with me, something with some weight so I can throw my rope over the branch. Tying my rope to the rock, I stand on the mound of a long-ago fallen tree to give me some extra height. It takes only a couple of throws to get the rope where I want it on the branch, and the almost empty pack pulls up easily. When it is out of reach of even the tallest of bears, I tie the rope to the trunk of a neighboring tree. There are places where the bears have had so much contact with humans that they have figured out that if they break the rope where it is tied to the tree, the pack will fall. Fortunately, Algonquin Park's bear population hasn't figured this out yet.

When I get back to my tent, I can see the sparkle of raindrops on the otherwise still lake. It isn't raining hard enough to feel it in the forest, and it's peaceful looking out over the quiet lake. A lone gull flies slowly and effortlessly in front of the site, eventually dipping down into the water to pick up a small fish. The herring gull is a beautiful bird. Unpopular in the city, gulls in their natural habitat are both graceful and stately. I once did some fieldwork at a herring gull colony and had the privilege of watching these birds raise their young. At four or five in the morning, we would walk through the colony to a blind that was set up amongst their nests. Although only one of us would stay in the blind, someone would have to go with them to fool the gulls. If only one person walked over and got into the blind, the gulls would never settle, knowing that something was up. But if two people walked to the blind and when one went inside, the other person walked back out of the colony, the gulls wouldn't realize that someone had been left behind. Some birds are more difficult to fool than this. To study ravens in Great Britain, five lucky researchers must drag themselves out of bed at four in the morning, just to leave one in the blind. Apparently, the ravens can count to four. If fewer than five people go to the blind, they know that you've left someone behind.

The day turns to dusk, and although the rain has stopped, clouds still obscure the setting sun. I'm hopeful that the sky will clear tonight and reveal the billions of stars that I can't usually see through the city lights. Or perhaps particles from the sun will dance on the waves of the Earth’s magnetic field, creating the ghostly lights of the aurora borealis. Before I get comfortable inside my sleeping bag, I drink down an almost full litre of water to ensure my “biological clock” will wake me up in the middle of the night. My sleeping bag is warm, and with the echoing call of loons from the lake, it doesn’t take me long to fall into a deep sleep.

Beyond Our Shores

My wake up call works as planned, and it is not much past one o'clock when I'm awake again with very little choice about going outside to check the night sky. I'm not disappointed. Although there is no sign of the northern lights, the sky is thick with stars. There are still a few clouds in the sky, recognizable only as the lack of stars in some places. I grab my paddle and flip the canoe over. The lake is quiet when I push off into the still water. From just ten or fifteen metres from shore, I can see the entire night sky. After getting comfortable on the bottom of the canoe, I almost instinctively search for the Big Dipper. Our evolutionary history has designed us so that we always look for patterns in the reality that surrounds us. Finding patterns allows us to build up a knowledge base by categorizing similar patterns and remembering them, rather than just living in a world of chaotic sights and sounds. As humans have become self-conscious, intelligent beings, we are sometimes misled by this instinct and find significance in patterns where none exists. Of course the stars in the constellations don't necessarily have anything to do with each other. Why should they? Stars that appear to be next to one another may actually be further from one another than they are from us. Looked at from any other vantage point in the galaxy, their patterns would be totally different. Yet we still find astrology columns in our newspapers as though the arbitrary patterns that these huge nuclear furnaces trillions of miles away make somehow determine if we are friendly or intelligent.

The patterns we find in the sky do, however, allow us to organize the stars in a way that we can remember. For this, the Big Dipper is a great starting point. If you follow the arc of the handle of the Big Dipper and continue beyond it following that trajectory, you will arrive at Arcturus, the fourth brightest star in the sky. Continue along this arc and you will come to Spica in the constellation Virgo. The rule to remember this is, "Arc to Arcturus and continue until you spy Spica". Virgo may not be the most spectacular-looking constellation, but I love to look into this region of the sky where the unimaginable distances of our universe reveal themselves. The central star in Virgo, Porrima, which appears much dimmer than Spica, is "only" about 35 light years away. With a small telescope, you can see that Porrima is actually two stars that are orbiting one another. Spica, on the other hand, is a single star, but it is hundreds of light years from Earth. It only appears so luminous to us because it is thousands of times brighter than our sun. But even Spica is still a near neighbor, contained well within the confines of our own galaxy. The universe outside of our own galaxy is revealed between Virgo's upraised arm and her head, where the Virgo cluster lies. Through a small telescope, this cluster appears not as discrete points the way that most stars do, but instead as a collection of fuzzy-looking points, which are given the name of celestial blurs. These particular celestial blurs are part of a huge cluster of over 3000 galaxies that are between 40 and 50 million light-years from Earth. It is mind-boggling to try to comprehend the distances and numbers involved. A single light-year is 9.46 trillion kilometers, so these galaxies are about half a billion trillion kilometers away (500,000,000,000,000,000,000 km) and each of these thousands of galaxies in this little cluster contains tens or hundreds of billions of stars. I don't think that there is any way to truly appreciate these sorts of numbers. Staring into the night sky, I am unable to put it into some perspective that I can understand. One thing is clear, though. Our tiny little planet is completely irrelevant, and with so many trillions of potential solar systems out there, it is not very likely that mine is the only species capable of looking into space and contemplating existence.

I shift my gaze back to the Big Dipper and follow the arc of the handle in the other direction, back to the two stars that make up the near side of the ladle. Following an imaginary line from these two stars through the bottom of the ladle brings me to Regulus. Regulus is the brightest star in the constellation Leo. To us, Leo still represents a lion, although the pattern is not quite as convincing as that of the Big Dipper. This particular pattern of stars was identified as a lion thousands of years ago, and for thousands of years through many different cultures Leo has remained as a constellation that represents a lion. Through time, different cultures have borrowed from the past, keeping the image of the lion but giving this constellation importance by incorporating it into their own mythologies. To the

Egyptians, the constellation was worshiped because the annual rise of the Nile coincided with the Sun's entrance into Leo. For the Greeks, it was said to have originated in the Moon, and its conquest by Hercules constituted one of the Twelve Labors. To the ancient Hebrews, it represented the traditional tribal sign of Judah. For medieval Christians, Leo was a symbol of the Lion's Den of the Book of Daniel.

The importance of the constellations to past cultures induced them to record the movement of the stars. The star Regulus was recorded as far back as 2100BC on ancient Babylonian tablets. It was by studying these tablets 2000 years after they were made that the Greek astronomer Hipparchus determined that the position of the Earth relative to the stars changes over time. This process, called precession, occurs because the slow wobble of the Earth's axis makes the whole night sky appear to shift over the course of thousands of years.

Leaving Regulus and following the two inner stars of the ladle through the top of the dipper leads to the star Deneb at the top of the Northern Cross. Although only the 19th brightest star in the sky, Deneb is one of the biggest supergiant stars known. Its diameter is about sixty times as big as our sun, and it's actually 60,000 times brighter than the sun. It only appears as dim as it does because it is 1600 light-years from our solar system.

By using the Big Dipper as a starting point, I am able to explore much of the night sky. But of course the most practical use of the Big Dipper is to use it to find the North Star, Polaris. The two stars on the far side of the ladle of the Big Dipper point almost, but not quite, directly to Polaris. Modern-day navigators and canoeists lost in the woods are fortunate to have such an easily located star pointing the way north. Because our friend Hipparchus recognized that the axis of the earth wobbles, we know that the more difficult to find star Thuban in the constellation of Draco would have been the north star at the time that the Egyptians were building the pyramids. In a thousand years, if we have any forests left, we will have to rely on some other method of finding our way out of the woods.

Climbing back into my sleeping bag feels particularly sweet after the chill of the night air. It seems I've only just closed my eyes when I am awakened by the sounds of morning. This morning, those are the sounds of the rummaging around and fluttering of wings of what must be Canada Jays. When I finally stick my head out of the tent, however, they have gone.

Though still early in the morning, I've finished a bowl of oatmeal and am all packed up and ready to head out. It's very calm and quiet, but the sky has become overcast again. It's cold out, and before I take a last look around the site, I pull my toque on. I dig a plastic plate that someone has left behind out of the fire pit and strap it onto the side of my pack. Otherwise, the site looks good. I plan to paddle back the way I came, but will continue along the little river past the first portage to do some exploring first.

It's nice to be in the canoe again, and it's always a pleasure to paddle on a perfectly calm lake. I paddle across the still water using the long sweeping J-stroke. I don't normally use this now fashionable steering stroke, but prefer the much maligned Goon stroke, or to use the less derogatory name, Superior stroke. The J-stroke is often touted as being more efficient than the Superior stroke, and from the point of view of the paddle this is probably correct. For those of you not already indoctrinated on how to do the J-stroke, the side of the paddle that pulls against the water on a regular stroke is turned away from the canoe at the end of the stroke. This contrasts with the Superior stroke, where the opposite side of the paddle is pried away from the canoe at the end of the stroke. So, from the point of view of the paddle, the J-stroke is probably more efficient. But from the paddlers' point of view, I don't believe that this is the case. The J-stroke is a far less powerful stroke than the Superior stroke, a fact that makes it useless in whitewater. When flatwater paddling, because it is less powerful, you must hold it longer at the end of your stroke than if you use the Superior stroke, and as a consequence, each paddle stroke takes longer to perform and much of your stroking time is spent steering the canoe rather than efficiently propelling the canoe forward. I think this more than offsets the argument that the Superior stroke slows the canoe slightly at the end of each stroke. In fact, after a reasonably powerful paddle stroke, you will see that there is a sort of whirlpool filled with air behind your paddle. When you turn your paddle to execute the Superior stroke, it meets with little resistance until reaching the far side of this whirlpool where it then quickly realigns the canoe to a straight course. Another inefficiency of the J-stroke is that it uses much smaller muscle groups than the Superior stroke. The Superior stroke allows you to use the muscles of your whole arm and shoulder, making it much easier on your body at the end of a long day's paddling.

Nonetheless, although it may be less efficient, the J-stroke is a graceful way to paddle a canoe. Long, leisurely paddle strokes seem more in keeping with the calm of the early morning water. The lake is quiet, and I turn the J-stroke into the perhaps inappropriately named Indian stroke, where the paddle never leaves the water, and try not to disrupt the silence.

Convergent Evolution: Adaptations of Alders

Paddling past my eventual portage, the little river meanders through a big open field of marsh. The marsh is made up largely of grasses, mostly brown now, with the new green shoots just coming into sight. There are also huge numbers of *Myrica gale* and leatherleaf amongst the grasses. As is often the case in marshy areas, sections along the edge of the river are lined with speckled alders (*Alnus rugosa*). Like the *Myrica gale*, the alders also obtain their nitrogen through a symbiotic relationship with the bacterium *Frankia* living in nodules in their roots.

Alders are monoecious, having separate male and female flowers on the same individual. The female flower, which looks distinctly like a cone, appears to make the alders unique among the broadleaved trees. These “cones” are actually woody catkins, however, and only superficially look like cones. Alders are members of the birch family (Betulaceae), and when looked at closely, the similarity between the “cones” of alders and the catkins of birch trees is quite clear. The male flower clusters are even more birch-like and don’t resemble anything found on the conifers.

If two species of tree are related, such as alders and birches, the fact that they both have a common distinguishing trait, catkin-like flowers for example, is because they share a common ancestor. This common ancestor would have been a species unlike either of the species alive today. Because of selection pressures at the time, individuals of this original species who had flowers that tended toward what we would now call catkin-like, would have left more descendants than those that didn't. In time, all individuals of this species would have had catkin-like flowers, as those individuals who had them gradually replaced those who didn't.

Of course, the fact that at some point all the individuals of this species shared specialized catkin-like flowers does not mean that they no longer faced selection pressures. This species would have continued to change over time, as all species do. At some point, some individuals of this species must have faced different selection pressures than others. As they all continued to adapt in response to their own environments, two groups of individuals must have begun to differ from one another. In one population, selection pressures favoured flowers that resembled what we now call true catkins; in another, selection pressures favoured cone-like catkins. The stage along this process that we see today is birch trees, with what we now define as a prototypical catkin, as well as alders, with their cone-like catkins. We can see that although they have the same unique flower type in common, this flower type has undergone different changes in each of these species. The alders and the birches of today have now become so distinct from one another that they can no longer interbreed, and so we no longer consider them as just separate populations, but instead define them as separate species. When populations or species increasingly diverge like this from their ancestral state, we call this divergent evolution.

In contrast to this is a case like that when we compare the cones of the alders with those of the conifers. By carefully comparing the cones, as well as the other features of each species, it's clear that the alders are not closely related to any of the conifers. The alders have independently evolved a female flower that is cone-like. Biologists call this convergent evolution; that is, when two species have evolved similar structures independently from one another. This is not particularly uncommon in nature, and the reason is quite simple. The vast array of species on Earth are undergoing constant change, and those whose changes help them survive are selected for, so when there is a good solution to a problem, more than one species is likely to find it. This is the case with cones. Cones are well designed to catch pollen, which is of paramount importance for wind-pollinated species like the conifers and alders.

Wind tunnel analyses of cones show that their geometry enhances the probability of catching pollen, and then directs that pollen towards the ovules. More than this, the cones are designed to preferentially entrap pollen from the same species of tree. Both conifers and alders have found this similar solution to the problem of catching the wind-borne pollen that their male flowers release.

Wind-pollinated trees have behavioral adaptations, which they have arrived at independently as well. Like many of the herbaceous plants in the forest, alders release their pollen early in the spring, while the trees are still bare, so that their pollen won't be intercepted by foliage. As well, the pollen is released only when the weather is clear and dry. If it was damp or rainy, the pollen would be washed out of the air and never make it to the ovules of a female. In fact, the pollen of most wind-pollinated species is usually released at a particular time of day. For alders, birches and pines, it is at about 2 PM, when the temperature is at its highest, the air is at its driest, and there is often a light breeze. If you suffer from allergies and notice that certain times of day are worse than others, it could be that the species of plants that you react to have a preference for releasing their pollen at that time of day.

Why Cooperate? Social Insects

Hanging from the bare branch of one of the cone-laden alder trees along shore is a large wasp nest. The grey volleyball-sized orb stands out on the leafless branch. This large paper nest will have been built the previous summer by a colony of social wasps from the family Vespidae. Almost none of the wasps that lived here last summer will still be alive. A few young queens will be the only members of the colony that will have survived over the winter. These surviving queens would have mated the previous fall and found a protected place to spend the winter. This large nest, so prominent along the shore, has fulfilled its purpose and has been abandoned. Other insects may take refuge in it for the time being, but it will soon fall apart.

In the early spring, each of the queens from this nest will find a suitable branch and will gather fragments of dead wood and vegetable fibers, which she will chew into a pulp to construct the first comb of cells. She will then lay her first eggs in this small comb of cells. When the larvae hatch, she will care for them diligently, catching and chewing up fresh insects to feed them. When these larvae mature into worker females, they will take over both the building of the nest and the collection of food for the increasing number of larvae. Once the first worker females begin taking over her duties, the queen will rarely leave the nest and will focus instead on producing more female workers. Later in the summer, the queen will stop producing female workers and will lay unfertilized eggs, which will develop into males. She will also lay some eggs in extra-large cells in the nest. These will become the new queens, who will mate with the males and then shelter for the winter so that they can start new colonies the next spring, starting the process again.

Because of their cooperative spirit, these wasps, and other social insects like them, have always been a source of fascination. The worker females, who make up the bulk of the colony, will work for the good of the colony and are even willing to defend the nest to the death, all without any hope of having offspring of their own. This behavior seems contrary to the idea of natural selection, where your only legacy is the descendants you leave in the next generation. Why do they do it? How could natural selection have produced what appears to be altruistic individuals who leave no offspring of their own, but instead seem only interested in helping others? As with most seemingly paradoxical situations, the reasons make perfect sense once we have managed to figure them out. The secret is that it is not specifically descendants that you want; it is rather your genes that you want carried into the next generation. Genes are what selection is actually acting on.

When humans, and most other animals, do what they can to ensure that their offspring are successful, it is because their offspring carry half of their genes. A blue jay has vast arrays of genes that give it the skills and inclination to build a nest, find a good mate, incubate eggs, and feed its young chicks. It acquired the genes that code for these behaviors from its parents. These genes survive from generation to generation precisely because they give the individual the skills necessary for those very genes to survive. If there were a blue jay who had a gene that gave it the inclination not to feed its young, its young would die. When that blue jay died, the gene that gave it the inclination not to feed its young would disappear. So we shouldn't be surprised that the genes that we see in most of the life around us code for behaviors that promote leaving successful offspring behind. These are in fact the only genes that can be the driving force behind each individual's existence; any genes that run counter to this just won't be around for very long. So your real descendants are your genes, they just happen to most commonly be packaged in our young. This is true of most sexually reproducing organisms, and specifically in humans, where we tend to draw our inferences from. Our children obtain half of their genes from their mother and half from their father, giving both the mother and father a large incentive to have children and to promote the survival of those genes inside of them. Or more accurately, both mother and father will have genes that give them the inclination to have and to care for their offspring because it is those genes in their parents that ensured that they survived.

The wasps and other social insects have slightly different incentives because they inherit sex in a very different way from us. A queen will lay two types of eggs: those that have been fertilized by a male, which will become daughters, and then some that have not been fertilized, which will become sons. Because her eggs that will become sons have not been fertilized, they contain only one set of genes, which are of course all from the mother. Her daughters, however, will have two sets of genes, one from their mother and one from their father. Usually with sexual reproduction, the daughters get a more or less random assortment of their father's two sets of genes as well as a random assortment of their mother's two sets of genes. With the wasps, however, daughters will receive a random assortment of their mother's two sets of genes, but all the daughters will share exactly the same genes from their father, since he only had one set to give. This puts each of the daughters in the unusual position of sharing half of their mother's genes, but all of their father's genes, with each of their sisters. So sisters will have ¾ of their genes in common. If one of these daughter females were to mate and have daughters herself, her daughters would share only half of her genes. To ensure the maximum number of their genes will be passed on to future generations, it makes more sense for daughters to help their mother to produce more sisters who will share ¾ of their genes, than it does for them to have offspring of their own who would only share half of their genes. The female workers in the hive aren't acting altruistically at all. They are doing what they can to promote the survival of their mother so that she can produce more sisters for them, allowing their genes to be propagated most efficiently.

Another Home Builder: Beavers

I don't paddle far beyond the abandoned paper home of the Vespidae before coming to a beaver lodge. Although made from unprocessed wood, a beaver's home-building skills are no less impressive than the wasp's. Unlike the wasps, the beaver's home has likely been used for many seasons. In fact, the beavers that currently live in this lodge may not have built it themselves, but instead may have inherited it from one of their parents.

There are lots of little trails from the river's edge that lead through the underbrush on shore, implying that there are likely beavers still living in this lodge. Where the water level doesn't fluctuate much, beavers will actually dig an upward sloping tunnel into the riverbank and build a subterranean chamber about a metre in diameter and half a metre high. They will cover the chamber with sticks and soil, but will actually live below the ground.

If the water level rises, they will scrape soil off of the ceiling to build up the floor, and will add more sticks and soil to the top. Sometimes they may end up being above ground and living in the lodge proper, but often times not. Where the water levels fluctuate, or tunnel building isn't possible, they will pile up sticks and earth away from shore and hollow out a living chamber inside of this. The important thing for beavers is to always have quick access to the water. They are not particularly well suited to life on the land, and can usually only avoid predators by getting to the water. In fact, coming to the New World the early Europeans convinced the Roman Catholic Church to regard beavers as fish by arguing that because of their impressive swimming ability and the fact that they had what look like scales on their tail, it should be permitted to consume them during Lent.

Traps and Turions: Bladderworts

In the middle of this little river, the water is dark, and I can thrust my paddle downwards, completely submerging it without touching bottom. I don't think the beavers need to build a dam here. The river itself doesn't have very much vegetation in it yet. Next to shore, I find a shriveled grey tangle of feathery leaves. This is a bladderwort (*Utricularia vulgaris)* from last year. In the summer, these feathery floating mats send up short shoots where several pretty little yellow flowers will bloom. It is a truly floating plant. It has no roots and instead spends its time floating just below the surface of the water.

As we have seen, most plants in the marsh have a difficult time getting nutrients even when they have roots. So perhaps it isn't surprising that, like many of the marsh plants, bladderwort captures and eats insects. In order to do this, they use what I think is the most impressive feat of engineering of all the marsh plants. Scattered throughout their feathery leaves are hundreds of little bladders, each one designed to trap and digest unsuspecting insects that venture too close. The bladders are like tiny balls with a door at one end. The closed door forms a perfect seal that doesn't let water either in or out. When the door is closed, the bladder actually forms a kind of vacuum by pumping the water out of the cavity until it is empty. When an insect, such as the larvae of a mosquito, brushes against the guard hairs that surround the doorway, the seal that is holding the door closed breaks. The water outside will then immediately rush in to fill the empty space inside the bladder, bringing the insect inside with it. The door seals closed again, and the bladder secretes digestive juices into the cavity until it has absorbed its prey. An unenviable death for the insect. The bladder will pump out the water again, resetting the trap.

The limp grey leaves that I now hold in my hand aren't in any shape to be capturing insects anymore. This year's plants will grow from special structures called turions, which would have been produced last season. At the end of the summer, the bladderwort produces these winter resting buds at the tips of its lower stems. They are basically an extension of the stem with a tightly compacted series of leaves that are wrapped into a tight ball. As winter comes and the stems begin to die, the turions break off the plant and sink to the bottom to wait out the winter. In the spring, as the water warms and the light of the lengthening days penetrates to the bottom, the turions activate. Their increased respiration and photosynthesis cause them to float to the surface where they sprout and begin the new season's growth. Much like how the trees determine when to open their springtime buds, the dormancy period of the turions is determined by following a set pattern and then adjusting it to account for external conditions. First and foremost, they have adapted to remain dormant for a certain length of time regardless of external conditions. After this time, they will determine when to activate based solely on external conditions. If it is still cold and dark, they will stay dormant longer. However, if conditions become favourable after the minimum dormancy time has passed, they will activate and rise to the surface.

A Lush Portage

Although the river continues, I paddle the canoe up to a well-worn path in the marsh that leads to a portage. A light drizzle falls as I make my way from log to log through the marshy path to the protective cover of the trees. I don't plan to continue on in this direction, but I want to explore the trail before heading back. The forest here is moist and lush. Unlike the other portages, the ground is thick with moss. Much like the last portage I explored, however, the only plants arising from the thick moss carpet are last year's *Cornus canadensis* (bunchberry) with their limp yellowed leaves, and the small seedlings of balsam fir. The forest floor here has even more deadfall scattered about than on the previous portage. The trees themselves are almost all coniferous. balsam fir, spruce, hemlock and white pine, all of different ages, give the forest a rich feeling. Many of the trees look as though they belong in a southern bayou, their dead branches draped with what looks like Spanish moss. This is Canada though, and on closer inspection this Spanish moss turns out to be the lichen known as old man's beard (there are several species from the genus *Usnea*). Like the reindeer moss (*Cladina rangiferina*) in the north, which is also a lichen, old man's beard often grows in bushy clusters, becoming abundant where it manages to get established. Oddly enough, the Spanish moss (*Tillandsia usneoides*), which grows in the south, isn't a moss either. It's a regular flowering plant that is in fact a member of the pineapple family, Bromeliaceae.

Old man's beard is one of the most commercially important lichens there are. Like many lichens, it produces its own unique acid. Called usnic acid, it protects the lichen by being an effective antibiotic against certain bacteria. People have learned to extract this acid from old man's beard, and it's used in ointments for skin infections and is also put into some deodorants to control bacteria. Millions of dollars are spent in the international market for this delightful fungal symbiont.

Underneath the green curtains of moss, there are many piles of brown oblong pellets a couple of centimeters long, indicating that this forest was well used by moose this winter. The scat left by moose and deer varies with the moisture content of their diet. The piles of distinct brown pellets commonly found throughout the forest are typically remnants from the winter when the deer and moose have had to survive on a woody diet. These pellets will hold their shape and may still be around at the end of the summer. During the summer, because of the high moisture content of the herbaceous plants, their scats will be loose and formless.

At the end of the portage are some large white pines, their cones littering the forest floor. Among these giant pines is a dead standing tree riddled with woodpecker holes. The woodpeckers take advantage of the abundant life inside the dead tree and can only survive because of the existence of this "dead" wood. The holes in this tree are quite large and almost rectangular, probably made by a pileated woodpecker looking for food. This is the bird whose call screams out through the forest, sounding more like a monkey in the jungle than a bird. There is no large nesting cavity in this tree, so the woodpecker that made the holes must live elsewhere. Most holes pileated woodpeckers pound in trees are to search for insects, usually carpenter ants, which they will pull out with their long sticky tongues. They will also chip large holes out of trees to use as nesting cavities, where they will lay four or five eggs that both parents will incubate. Often woodpeckers will pound on trees, not for the purpose of chipping out holes, but to attract mates or claim territory. This is why you can sometimes find woodpeckers pounding their beaks against metal roofs or siding.

From the end of the portage, a creek, smaller than the one that I came from, meanders through the low flat marsh into the distance as far as I can see. The marsh is not wide, and it ends abruptly where it meets a thick coniferous forest on either side. Looking over the flat marsh, the forest on each side frames a large hill in the distance, almost convincing me to bring the canoe across the portage and explore further. The time I have won't allow it though, so I reluctantly turn back.

Whose Genes are in Charge? Plant Galls

Heading back along the trail, the ground is soft underfoot. It's raining harder, and the trees no longer keep it from reaching the forest floor. On a small branch of a spruce tree, I notice a gall. This bulbous woody protrusion looks unnaturally placed, and it seems odd that a perfectly normal-looking stem would suddenly decide to grow a golf ball sized sphere on its side. Really, the only odd thing about the gall is its shape. The colour and texture look the same as the rest of the stem, and it was clearly made by the tree itself. Galls appear on most species of plants and have a huge variety of forms and causes. They may be like this bulbous outgrowth on a stem or branch, or they may be a strange growth on a leaf or bud. Even most witches' brooms, the bushy pockets of stems and foliage that look a bit like birds' nests, are a kind of gall.

Galls are caused when another organism infects the plant. This could be viruses, bacteria, fungi, nematodes, or insects. In many cases, the formation of galls is a response of the host plant. The plant produces the gall in order to isolate the invader and protect itself from further infection. Producing a gall does not normally use a significant amount of resources, and so has only a minor effect on the health of the plant. For the invading organism however, although it has been denied access to the rest of the plant, the invader can usually benefit from this new place to live, and has adapted over time to take advantage of the new conditions created by the plant.

Many other galls, on the other hand, are produced under the direction of the invading organisms. The plant is not just creating new layers of cells. Instead, the invader controls the growth and differentiation of these cells into structures that are completely foreign to the host plant. The invader can force the plant to use its own tissue to build a made-to-order structure where the invader can live or have its progeny protected well they develop. In some cases, specialized cells may be initiated in the center of the gall to give the invader or its progeny a continuous food source. These cells will not only be of a form that is not found in the plant, but will have a higher concentration of starch, sugars, lipids, and proteins. These gall inducers not only control the form, structure and colour of the gall produced, but also the movement and storage of nutrients in the plant. They successfully manage to create a home for themselves using the machinery and resources of the plant. Relationships like this force us to take a very different view of gene expression. It would be hard to guess that a gene that is firmly embedded in the DNA of an insect might code for the production of a certain kind of cell in a specific species of plant.

Because many of these galls have high concentrations of nutrients and offer protection, they are themselves a new habitat, so it's not surprising that other organisms have adapted to take advantage of this. Perhaps the most specialized of these are the fungi or insects that can't induce a gall of their own, but feed on gall tissue. The fungi may be an endophyte. As we have seen, the endophytic fungi sometimes take advantage of the fact that they are already present in the leaves of the plant to become decomposers once the leaves fall to the ground. In the same way, some species will take advantage of already being present in the plant and adapt to grow into the gall and feed on the gall tissue, sometimes killing its host. The insects that take advantage of galls are often wasps that will lay their eggs in a particular kind of gall where their larvae will become enclosed in their own chamber and will feed on the nutritious cells. Some species will kill the original gall producer, or will use the original gall producer itself as their food source. Many others will not harm them; after all, it is the gall inducer's offspring that will produce the new galls your descendants need to survive.

Insects such as these aren't the only ones to have adapted to take advantage of the new opportunities created by the gall. In fact, the very presence of these insects creates even more opportunities. Not only can their presence increase the size of the gall, but the gall now contains two species, which other organisms can take advantage of. Whole micro-communities can develop inside galls, including bacteria, fungi, nematodes, spiders and insects. The extent of these micro-communities can be amazing. Oak-apples, those unusual "fruit" that look like apples and sometimes grow on oak trees, are induced by a cynipid wasp. However, there are up to 75 different species associated with an oak-apple gall. The gall itself has become an entire community of interacting organisms, all taking advantage of the wasp's home.

I return to the canoe and eat a handful of nuts before heading back along the river. The water is still calm, but the rain hasn't let up. Meandering back through the marsh, past the homes of the beavers and wasps, I reach my portage. This was a short Class II, a nice walk over dry rocky high ground.

More Selection for Cooperation: The Ants

Pulling my canoe up onto the grass near the start of the portage, I notice the sandy mound of an ant nest. It's a little cool and wet today for there to be many ants around, but the dozens of little entrances make it clear that the ants have been busy. Ants can successfully live in colonies, as the wasps do, because they are also members of the Order Hymenoptera and have the same method of sex inheritance. So, some of the same selection pressures have been acting on the ants, making it in the young females' best genetic interest to help raise sisters rather than have offspring of their own. An ant queen will typically live five years or longer and, in fact, a queen of the European black sidewalk ant (*Lasius niger*) survived in a laboratory nest for 29 years, making it the longest living insect known.

Apart from the incredible cooperation shown by their ability to live in communal colonies, which we already talked about with the wasps, one of the most interesting things about ants is their relationship with aphids. If you search through the plants of almost any weedy field, you can find tiny aphids sucking the sap from them. If you watch these aphids for a few minutes, you will likely see a worker ant moving from aphid to aphid, licking the sugary liquid from the anus of each one. This is another symbiotic relationship that has evolved for the benefit of both species.

The aphids are in fact not sucking the sap from plants. Sugars are transported throughout plants in a series of tubes called the phloem. These phloem tubes are located close to the surface around the circumference of the stem. The sugars are forced through these tubes under pressure as the plant transports them from either their leaves to their roots, or their roots to their leaves, depending upon their needs. The aphid has adapted to take advantage of both the food that the phloem carries, and of the pressure that transports this food.

Aphids have developed long hollow mouthparts, which they can insert into the phloem. Inserting this hollow tube into the phloem is like sticking a siphon into the side of a fire hose. The liquid, which is full of not just sugars but also proteins, minerals and vitamins, is much more than the tiny aphid needs. So the aphids pass much of this liquid through their bodies and out of their anuses. It is to their advantage to tailor it to the needs of the ants if the ants can offer something in exchange. The ants do collect this freely available food source and protect the seemingly defenseless aphids from predators in exchange.

Some species go much further than this, and the ants will store the eggs of the aphids in their nests over the winter. When spring comes, the ants will carry the newly hatched nymphs to the roots of food plants. If these plants die, the ants will move them to another plant. The ants even take into account the aphids' stage of development and move them not only to the right species of plant but also to the part of the plant that the aphid needs at that time. The quiet anthill that I see may contain not only thousands of ants but also the nymphs of aphids waiting to be transported to their food source.

Walking through the forest amongst the trees, it is calm and not at all unpleasant. The tops of the trees tell a different story, however. The wind is picking up, and the skies are dark. I hurry across the portage, mildly concerned that if the weather gets worse, I might have to do an unpleasant lake crossing to get back.

Attack of the Slime Mold

The next portage is only metres away, and although I've picked up the pace a bit, I'm not far along its path before I notice a dull yellow mass on the side of a fallen log. It looks to be the slime mold *Fuligo septica*, and I can't pass up having a closer look. Slime molds have got to be one of the coolest creatures to be found in the forest. For a long time they were considered to be fungi and are called molds because they look so much like some sort of fungal growth, and like fungi, they even reproduce by forming spores. However, the lives of slime molds are more interesting than the average fungus. The yellowish fungal-looking mass is actually a plasmodium, a blob of protoplasm without cell walls, much like a giant amoeba. This blob actually crawls across the forest floor, up and over stumps and fallen trees, searching for food. As it crawls over bacteria or other microorganisms, it engulfs and digests them. When the environment dries out or food starts to run low, the slime mold will usually crawl up a tree trunk or rock where it will change from an amorphous blob and become more crust-like. This is a much more fungal seeming stage in its life-cycle where it remains in one place and becomes a spore-producing body. These spores are released from as high a point as possible so that the wind can carry them throughout the forest. If the spore lands in a suitable place, it will divide and produce little amoeba-like cells, which will fuse and form a new plasmodium.

These blobs of protoplasm can appear quite suddenly. In Dallas, Texas, when a giant blob almost a meter across appeared on a neighborhood lawn, fear of alien life-forms spread across the city. The fire department was dispatched, and when they tried to destroy it with a high-powered hose, it only made the blob grow. Fears intensified, and articles about the alien blob had come out in the national media before a local science student finally identified it as a slime mold.

But perhaps we shouldn't be so confident that the slime molds aren't planning world domination. Scientists have performed tests in the laboratory to determine the "intelligence" of slime molds. Researchers set up a maze with food at the start point and an end point but not in between or at any of the dead ends. It took only eight hours of searching through the maze, changing course and backtracking several times for the plasmodium to find the shortest route between the two points. Once the plasmodium found the shortest route, it stopped searching the other avenues and formed its body between the two points with the food. Not bad for an oozing pile of protoplasm.

Heading Home

Back in the canoe, safely beyond the reach of the slime mold, I stop to pump a bottle full of drinking water. I'm fairly confident that I could drink the water without bothering to pump it, but I'm not immune to the over-cautious approach of our times and so I diligently pump every drop I plan to drink. When I'm on a several week long canoe trip down a river, I often think about the fact that I am mostly made up of water. It seems strange that something as solid and of such a precise shape as my body could be mostly liquid, but water makes up about 65% of the weight of my body. It's remarkable to think that after paddling for many days, drinking nothing but water from the river, the liquid in my body will gradually replace itself, and in the last days of the trip, the majority of my body will be made up of that river. In a purely physical sense, you do become one with the river.

The skies are still dark, but the water isn't too rough when I paddle into the large expanse of the lake. The big lake still has that wintry look, and I strain to notice if the poplar leaves have opened just a little bit more since yesterday. I can't help reflecting that the next time I come here it will seem like an entirely different place. The trees will be thick with leaves trying to harness as much of the summer sun's energy as they can before winter comes around again. Shaded from the sun, the spring flowers will have died back, and the already inconspicuous fungi and mosses will be even harder to find as the long hot days become drier. The most significant change, however, will be the huge numbers of insects that will suddenly appear on the scene. Not the least of which will be the millions of blackflies and mosquitoes that we all love. Dragonflies will battle for territory along the lake's shores, caterpillars will hang from silky threads from the trees above, fireflies and moths will bring the nighttime to life, and beetles will clumsily fly between the flowers of the water lilies. By just sheer weight of numbers, the insects will bring the forest to life, not just with their own hums and buzzes, but by keeping the chirping frogs and singing birds fed.

For today, the wide, open water marks the end of my meanderings. The far side of the lake is a clear destination, and it is invigorating to paddle hard through the waves. The richness and complexity I have witnessed over the last couple of days has been, as always, both humbling and inspiring. Species and landscapes are constantly changing as they adapt and evolve, but nature, indifferent to the changes, continues as always. The fact that every species on earth is ephemeral, existing for an insignificant portion of the earth's time, makes each of them all the more wonderful to watch and try to understand. But it is the process that is beautiful, and I like the fact that I am unnecessary in the process. Nature moves along without me.

As I approach the shore and am taking my last paddle strokes, I see a herring gull ripping apart a fish on the shore. After a weekend observing all the subtleties of cooperation and competition in nature, perhaps it's fitting that the trip should end with such a classic, straightforward relationship between predator and prey.

Acknowledgments

After publishing my first novel, *HUMAN*, in 2025, I decided to re-release *Paddling with a Naturalist*, the very book you now have in your hand (or on a server somewhere). *Paddling with a Naturalist* was originally published in 2010 when the publishing landscape was very different, and as well as having made a fair number of edits and improvements to this version, I'm also able to offer it at a lower price.

Having just recently written *HUMAN*, it couldn't escape my notice that it's the very same people who I would like to acknowledge for making *Paddling with a Naturalist* possible. So please forgive me if I shamelessly plagiarize much of what's below from myself.

When I was a child, my parents imparted to me a love of reading, but I think what was more important is that they encouraged me to question everything and to be curious about how and why things are the way they are. In particular, they encouraged an interest in the natural world. I came to understand, eventually, that not everyone's parents drove them around looking for roadkill to bring home to boil the flesh off the bones. Nor did everyone's parents allow them to catch and keep snakes over the summer, releasing them when fall came. Seventeen snakes in the house was, I recall, the most successful year. This despite my mom's not fear, but uneasiness about snakes. As she noted, with her characteristic wit, she didn't really mind the snakes, it was just their tails she didn't like. So, before all, I would like to thank my parents, Pat and Guy Hodnett, for beginning the journey which brought about this book.

I would also like to thank my multi-talented son, Ryan Hodnett. First, for his direct assistance with the creation of this book, without which, the process would have been much more painful and fraught. Specifically, I would like to thank him for the creation of my website, the majority of the design and creation of the book cover, and the formatting of *Paddling with a Naturalist* into an EPUB. For that, as well as for the inspiration I derive from how Ryan approaches the world, and simply for his existence in it, I am also grateful.

Finally, I would like to thank my wife, Karen Hébert. Sharing a life with someone sets you on a particular trajectory. For me, it has been a trajectory of love, laughter and discussion, along with a bit of shared outrage at the world. I am forever grateful to her for that. I am also grateful for her encouragement and patience when I first wrote this book, and her helpful suggestions were invaluable. For this and more, she has my everlasting love and thanks.

I'd also like to thank you for reading *Paddling with a Naturalist*. As an independent author, building visibility depends on personal recommendations. If you have the time and inclination, please consider leaving a short review wherever possible.

www.ingramcontent.com/pod-product-compliance
Lightning Source LLC
LaVergne TN
LVHW022222200926
848481LV00006B/2075
9781068907029